Lecture Notes in Biomathematics

Managing Editor: S. Levin

W0079539

50

David H. Anderson

Compartmental Modeling and Tracer Kinetics

Springer-Verlag
Berlin Heidelberg New York Tokyo 1983

Author

Dr. David H. Anderson
Department of Mathematics
Southern Methodist University
Dallas, TX 75275, USA

ISBN 978-3-540-12303-3   ISBN 978-3-642-51861-4 (eBook)
DOI 10.1007/978-3-642-51861-4

2146/3140-543210

PREFACE

This monograph is concerned with mathematical aspects of compartmental analysis. In particular, linear models are closely analyzed since they are fully justifiable as an investigative tool in tracer experiments. The objective of the monograph is to bring the reader up to date on some of the current mathematical problems of interest in compartmental analysis. This is accomplished by reviewing mathematical developments in the literature, especially over the last 10-15 years, and by presenting some new thoughts and directions for future mathematical research.

These notes started as a series of lectures that I gave while visiting with the Division of Applied Mathematics, Brown University, 1979, and have developed into this collection of articles aimed at the reader with a beginning graduate level background in mathematics. The text can be used as a self-paced reading course. With this in mind, exercises have been appropriately placed throughout the notes. As an aid in reading the material, the end of a proof is indicated by $\varslash\!\!\!/\!\!\!/$. Subsection titles are utilized to make it easier for the reader to skim over detailed material on a first reading and make the entire manuscript somewhat more accessible, especially to nonmathematicians in the biosciences.

The preparation of this monograph has been a long task that would not have been completed without the influence of a number of individuals. I am especially indebted to H. T. Banks, J. W. Drane, J. Eisenfeld, J. A. Jacquez, D. J. Mishelevich, and A. S. Perelson. I am also grateful to the Department of Medical Computer Science of the University of Texas Health Science Center at Dallas, which provided much of the atmosphere and monetary resources to make this project possible. Special thanks go to my wife, Susan, and to Carolyn Simmons for their excellent typing of the notes.

David H. Anderson
February 1983

TABLE OF CONTENTS

SECTION 1. COMPARTMENTAL SYSTEMS

1A. Introduction

In this monograph we shall discuss the analysis of biological systems by means of compartmental modeling and tracer studies. This type of modeling has application in a variety of areas such as drug kinetics in pharmacology, studies of metabolic systems, analysis of ecosystems, and chemical reaction kinetics. The fundamental approach in this style of modeling is to analyze a system by separating it into a finite number of component parts, called compartments or states, which interact through the exchange of material. The mathematical theory for the behavior of these systems of compartments is called <u>compartmental analysis</u>, and it has been found to be quite useful in the analysis of experiments in many branches of the biosciences.

The application of compartmental analysis is like that of general mathematical modeling. One goal is to develop a <u>plausible mathematical representation</u> of a particular biological phenomenon. This model will be a simplification of that phenomenon, but it should be an adequate representation for the purpose in mind. It requires considerable depth of knowledge in the field from which the problem is taken to formulate a meaningful and useful mathematical model. This is usually best accomplished by finding someone knowledgeable in the area to provide the appropriate background and justification for the use of a compartmental model. Another goal is to develop and expand the <u>theory of compartmental systems</u>. In this book a considerable amount of space is devoted to structural properties of compartmental systems, especially in the linear case. A third goal in compartmental analysis is to deal effectively with the <u>inverse problem</u>, which consists of the three phases: model specification, structural identifiability, and parameter estimation. Here <u>model specification</u> refers to the determination of the number of compartments and the interconnections between compartments which allow the exchange of material. In the <u>structural identifiability</u> aspect of this problem, we inquire as to whether or not the system parameters are uniquely defined under error free observations, given that the model is completely specified including the compartments receiving inputs and those which are observed. We will observe that the experimental information found and its relation to the model parameters are given by a set of nonlinear algebraic equations, and as a consequence, the structural identification problem reduces to the question of whether or not a nonlinear algebraic system of equations possesses a unique solution. Provided the structural identification of the system obtains, the practical problem of computing the "best" values for these model parameters—— parameter estimation——still remains. We will discuss analytical and numerical methods of calculating these best values.

1B. Preliminary Definitions

Kinetics is that branch of dynamics that pertains to the turnover of specific particles in a biological system. A compartment is an amount of material which acts kinetically in a homogeneous distinct way. The compartment to which a particle belongs characterizes both its physical-chemical properties and its environment. A compartment may not be an actual physical volume; however, the amount of some material in a physiological space is often treated as a compartment in certain clinical studies. The particles of each compartment are influenced by forces which cause the particles to transfer from one compartment to another. All particles in a particular compartment have the same probability of transition since within the compartment all particles are well-mixed and considered indistinguishable by the system. The transition from one compartment to another occurs by passing through some physical barrier or by undergoing some physical or chemical transformation. Associated with each compartment is its size. Sometimes the terms volume and size are used interchangeably.

A compartmental system consists of two or more compartments, interconnected in the sense that among certain compartments there is exchange of material. The com- partmental system will be primarily modeled in a continuous deterministic manner by a collection of ordinary differential equations, each equation describing the time rate of change of amount of material in a particular compartment. These rates of change are dictated by the physical-chemical laws that govern the material exchange between interacting compartments, e.g., diffusion, temperature, chemical reactions, etc. It will be assumed that the set of equations model average properties of the

system for very large numbers of particles. From the idealized physical model it is
hypothesized as to the exact nature of the interconnections that occur between compartments. In the usual representation for a compartmental system, a box denotes a
compartment, and an arrow indicates the transfer of material into or out of a compartment (see Figure 1.1). There also may be inputs from the outside environment
into one or more compartments (vertical arrows pointing into the tops of boxes), and
there can be excretion of material from some of the compartments to the outside
environment (vertical arrows pointing out of the bottoms of boxes). Should there be
no exchange of material to the outside environment, the compartmental system is
referred to as closed; otherwise it is said to be an open system. Realistically,
many compartmental systems are open, for some material is lost by excretion,
metabolism, etc.

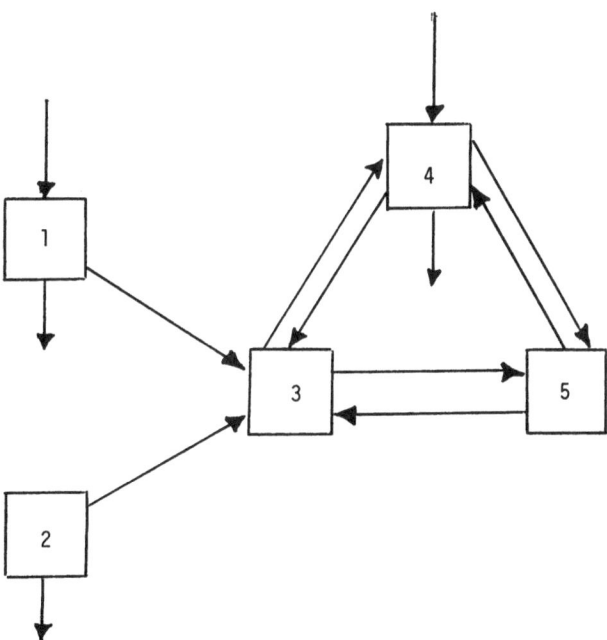

Figure 1.1

1C. Tracer Experiments

Higher-level organisms tend to maintain characteristics such as temperature
within fairly uniform ranges, even with slight disturbances that occasionally occur

within their systems. The maintenance of a <u>steady-state process</u> in a physiological system is often studied by means of compartmental modeling and <u>tracer experiments</u>. In such steady-state studies, the <u>fractional transfer coefficients</u> (which are associated with an exchange of material from one compartment to another, and may be labeled on each arrow of the compartmental system diagram), the <u>excretion coefficient</u> for the transfer of material to the outside, and the <u>sizes</u> of certain compartments, are usually unknown. To estimate these quantities, the following procedure has been developed (see Section 7). A known dose of <u>tracer</u> (also called <u>labeled</u> or <u>tagged</u> material) of negligible volume is injected into one or more compartments of the system, say at time $t = 0$. This introduction of labeled material at time zero perturbs the system in that a slight excess of material is suddenly present. However, the amount of tracer present is minute enough that it is reasonable to assume that the steady-state characteristics of the system remain unchanged. For $t \geq 0$, there are two types of particles in any compartment of the system--one labeled (the tracer) and one <u>unlabeled</u> (the original material in the compartment, sometimes called the <u>tracee</u> and <u>untagged</u> material). The two kinds of particles are assumed indistinguishable by the system, but can be distinguished by the observer since the tracer is a quantity of material such as a dye or a radioactive isotope. Due to the system interconnections, the tracer permeates the entire system, and is eventually drained from the system through any exits present.

When adding a tracer to a system in steady-state, the amount injected should be small enough so as to not disturb the steady-state processes and thus not adversely affect the functioning of the living organism.

The fact that compartments are assumed to be well-stirred ensures that each compartment is a homogeneous mixture of tracer and tracee. Thus it is assumed that the dynamic behavior of the tracer in a compartment is representative of the dynamic behavior of the tracee in that compartment. Moreover, homogeneity has the implication that upon entrance of tracer into a compartment, it is instantaneously mixed throughout the tracee.

The tracer is detectable in the system for only a temporary period of time after t = 0. During this transient period, measurements of the tracer are taken sequentially in time in at least one compartment. The objective of this sampling of the tracer is to obtain estimates of the parameters of the model and thereby get a plausible mathematical description of the system being studied. This is the inverse problem, and its solution gives information on the steady-state characteristics of the compartmental system.

Perhaps the tracer experiment is best understood if we think of a collection of tanks each containing an amount of water (tracee). Each tank is provided with a controllable inlet for water and an open or closed drain. Every pair of tanks is connected by two pipes through which a fixed amount of water is allowed to flow (could be a zero amount). Generally, the amount of flow in one direction is different than the other. Also each tank has a blender which thoroughly stirs the water in that tank. Each inlet flow of water can be adjusted so that the water exchanges between tanks are kept at steady-state, and so the volume of water in every tank is constant over time. Then at time t = 0, a small amount of dye (tracer) is dropped into one or more of the tanks. The blenders ensure that the tracer concentration is uniform in each tank of the system. In one or more of the tanks the tracer is then measured at discrete points over time with the goal being the determination of the flow rate of water through each pipe and drain, and the calculation of the water volume in each tank.

1D. History of Compartmental Analysis

The utilization of compartmental models is fairly recent. In 1948, Hevesy [127] wrote a book in which he demonstrated applications of radioactive tracers, with some of his work dating back more than two decades. He discusses how tracers can be used to determine the distribution and excretion of material in the body, as well as the preparation of tracers. In 1966, Rescigno and Segre [188] published a book (the Italian version appeared in 1961) which contains many specific cases of compartmental modeling. The Sheppard text [210] of 1962 addresses the development of more complex compartmental models and starts to consider general theories; for example, a complete chapter is devoted to the study of mammillary systems. Atkins [16] published his work in 1969 followed by Jacquez [134] in 1972. The book by Jacquez is now the standard reference on compartmental analysis. It covers the basics of linear and nonlinear compartmental systems, radioactive tracers, and the inverse problem. Applications, stochastic modeling, and control are also discussed.

SECTION 2. ELEMENTARY COMPARTMENTAL MODELS

In this and the next three sections, we shall present a variety of examples of compartmental systems, both linear and nonlinear. Each illustration fits under the category of a general compartmental model, as will be defined in Section 6.

Besides those examples given here, elementary compartmental models can be found in references [8], [19], [29], [34], [39], [68], [77], [130], [169], [190], [191], [195], [207], [215], [216], [218], [230], [235].

2A. Drug Kinetics

Our first compartmental example is the simple one of examining the ingestion and subsequent metabolism of a drug in a given individual (e.g., see [161] and related references [70], [163], [188]). We use a 2-compartment system, as illustrated in Figure 2.1, to model this process in the body.

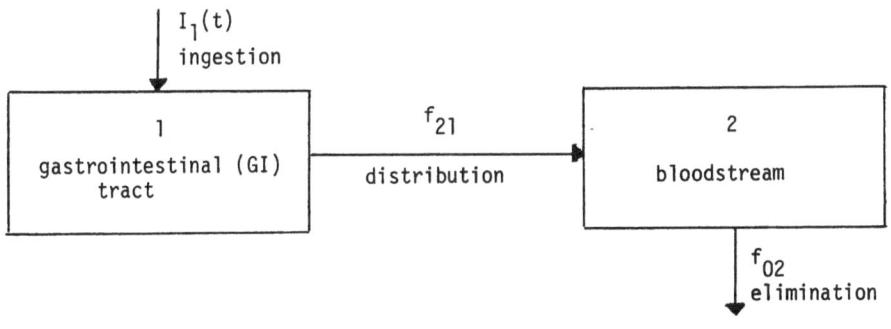

Figure 2.1

Suppose the drug is taken as oral medication. It then enters the GI tract, is absorbed into the circulation and distributed throughout the body to be metabolized and finally eliminated. Let time t start at zero. Let $q_1(t)$ denote the mass of the drug in compartment 1 (GI tract), and $q_2(t)$ be the mass of drug in compartment 2 (bloodstream). If the ingestion rate of the drug is $I_1(t)$, then a basic plausible assumption for a model is that

(2.1) $\dfrac{dq_1}{dt} = I_1(t)$ - drug distribution rate from compartment 1 to 2.

Equation (2.1) is commonly called a <u>mass balance equation</u>. In the simplest case of first-order kinetics, the latter rate in (2.1) is assumed to be proportional to the mass (or concentration) of drug in the GI tract. Let $f_{21} > 0$ be that proportionality constant. Then (2.1) becomes

(2.2) $\qquad \dot{q}_1 = I_1(t) - f_{21}q_1$.

For compartment 2, we also take

(2.3) $\qquad \dfrac{dq_2}{dt} = $ inflow rate - outflow rate

where $f_{21}q_1$ will be the inflow rate of drug distribution from the GI tract. In keeping with the above first-order kinetics assumption, we take the outflow rate from compartment 2 to be proportional to q_2. Thus (2.3) becomes

(2.4) $\qquad \dot{q}_2 = f_{21}q_1 - f_{02}q_2$

where $f_{02} > 0$ is this new constant of proportionality. In (2.4), observe that the instantaneous time rate of change of q_2 is given by the difference between two first-order processes. Equations (2.2) and (2.4) along with appropriate initial conditions $q_1(0)$, $q_2(0)$, constitute the drug metabolism model. In a matrix-vector format the constant coefficient linear differential equation model is ($t \geq 0$)

(2.5)
$$
\begin{bmatrix} \dot{q}_1(t) \\ \dot{q}_2(t) \end{bmatrix} = \begin{bmatrix} -f_{21} & 0 \\ f_{21} & -f_{02} \end{bmatrix} \begin{bmatrix} q_1(t) \\ q_2(t) \end{bmatrix} + \begin{bmatrix} I_1(t) \\ 0 \end{bmatrix} .
$$

We are especially interested in $q_2(t)$, the mass of drug in the bloodstream, since this variable probably tells us the most about how the drug will effect the person, and because that compartment is accessible to analysis through taking blood samples. One purpose of this study is to see how different values of $I_1(t)$, $q_1(0)$, $q_2(0)$, f_{21} and f_{02} affect the amounts of drug in the bloodstream.

2B. Leaky Fluid Tanks

There is an obvious analogy between the above model and the amounts of fluid in leaky tanks, as pictured in Figure 2.2. Input of fluid is made to tank 1 at a rate $I_1(t)$. This tank has an initial amount $q_1(0)$ of fluid, and a leak measured by the parameter $f_{21} > 0$. This leak allows fluid to empty into the second tank which has initial amount $q_2(0)$ of the same fluid, and another leak measured by parameter $f_{02} > 0$. Figure 2.2 illustrates an instantaneous view at some time $t_1 > 0$ of the 2-tank system where initially tank 1 was full and tank 2 empty. The same mathematical system as before, equation (2.5), can be used to model the dynamics of the fluid in the successive tanks. The main goal here, in analogy to the drug metabolism model, might be to know the quantity $q_2(t)$ of fluid in tank 2 at any positive time t.

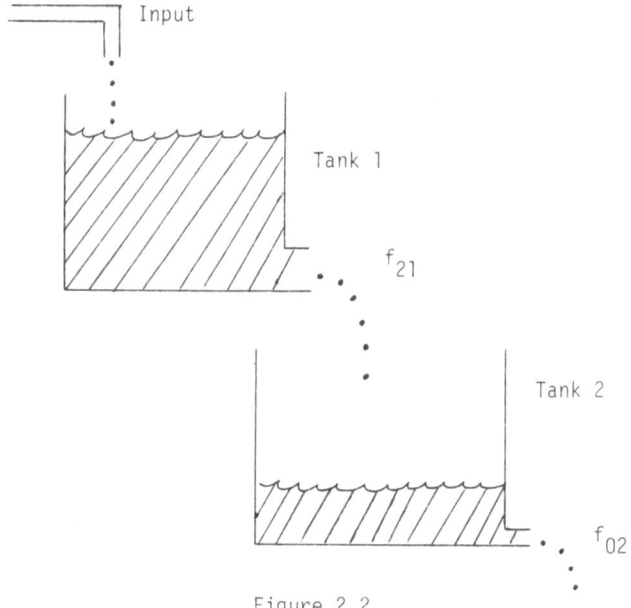

Figure 2.2

Exercise 2.1. In model (2.5), suppose $I_1(t) \equiv 0$. Find solutions for $q_1(t)$, $q_2(t)$. Graph $q_1(t)$, $q_2(t)$ when $f_{21} = 0.5$ and $f_{02} = 0.25$. Discuss the asymptotic behavior of q_1, q_2 as $t \to \infty$. For the model to make sense physically, are there any other constraints that must be put on f_{21} and f_{02}?

Exercise 2.2.   Suppose $q_1(0) = q_2(0) = 0$ and $I_1(t)$ is a nonzero input, say exponen-

tial: $I_1(t) = c \exp(-\rho t)$, $t \geq 0$, where we suppose $\rho > f_{21} > f_{02}$. How do the

solutions for q_1, q_2 differ from that of the previous exercise?

Model (2.5) is simple enough to be solved analytically, but the reader should

also solve the equations numerically (simulation, using graphic capabilities), and

in the process experiment with the effects of each of the individual parameters

$q_1(0)$, $q_2(0)$, f_{21}, and f_{02}. This develops insight about the role the parameters

play in determining the trajectory of the amount function $q_2(t)$. For example, among

other things, one can observe from either simulation or the analytic solution that

the maximum value attained by $q_2(t)$ when $I_1(t) \equiv 0$ depends upon the <u>relative</u> values

of f_{21} and f_{02}.

2C. Diffusion

Let us now consider the diffusion of a material between two compartments sepa-

rated by a membrane through which the material can diffuse, though not necessarily

with the same permeability in the two directions (see Figure 2.3).

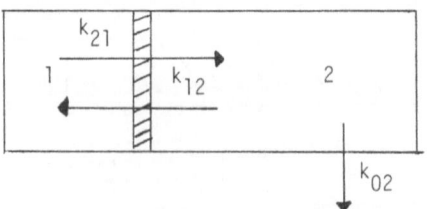

Figure 2.3

Let $c_1(t)$, $c_2(t)$ be the concentrations of the material in the two well-mixed

compartments of fixed volumes V_1, V_2, respectively. <u>Fick's Law</u> states that the rate

of material transfer across the plane of the membrane by diffusion is proportional

to the product of the membrane surface area A, and the concentration gradient between

the two regions. Since the compartmental volumes are constant, we can write $q_i(t) =$

$c_i(t) V_i$, $i = 1$, 2, for the mass of material in compartment i. Thus by Fick's Law

we can write the instantaneous rate at which material crosses the membrane from

compartment 1 to 2 as

$$(2.6) \qquad \frac{dq_1}{dt} = KA(c_2 - c_1)$$

for some positive proportionality constant K. Hence, from (2.6), the rate of exchange of material between the two compartments is the sum of two first-order processes: $- KAc_1$ and KAc_2. If in addition we assume that excretion from compartment 2 is at a rate proportional to the concentration of material in that compartment, then the principle demonstrated in (2.6) as applied to the total diffusion system of Figure 2.3 yields the model equations

$$(2.7) \qquad \begin{aligned} \dot{q}_1 &= - k_{21}Ac_1 + k_{12}Ac_2 \\ \dot{q}_2 &= k_{21}Ac_1 - k_{12}Ac_2 - k_{02}c_2, \end{aligned}$$

where we have allowed the permeability to be different in the two directions. Introduce the "fractional transfer coefficients"

$$f_{ij} \equiv k_{ij}A/V_j, \qquad f_{02} \equiv k_{02}/V_2$$

into equations (2.7) to get the diffusion model

$$(2.8) \qquad \begin{bmatrix} \dot{q}_1(t) \\ \dot{q}_2(t) \end{bmatrix} = \begin{bmatrix} - f_{21} & f_{12} \\ f_{21} & - f_{02} - f_{12} \end{bmatrix} \begin{bmatrix} q_1(t) \\ q_2(t) \end{bmatrix},$$

a system similar in structure to (2.5).

2D. Solute Mixture

The same type of equations as in (2.8) result from consideration of a mixing problem for two tanks each containing, say, a salt solution (e.g., see [161]). The well-stirred tanks and their interconnections are shown in Figure 2.4. It is assumed that each tank i remains full at constant volume V_i, and that there is a salt solution in each tank.

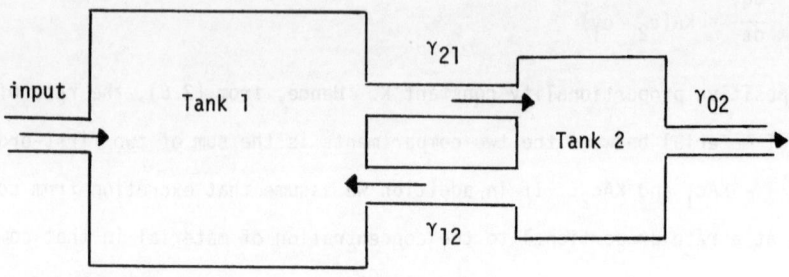

Figure 2.4

Suppose the solution in tank j is pumped to tank i at a constant flow rate γ_{ij} ($i \neq j$). Let q_i represent the mass of salt in tank i and c_i its concentration. Assume that the excretion of the salt from the second tank is proportional to the concentration of salt in the solution, with proportionality constant γ_{02}. If the total rate of change of mass of salt in a tank is the rate at which salt is added to that tank minus the rate at which salt is withdrawn from that tank, then a plausible model for the dynamics of the salt in solution is

$$\dot{q}_1 = \gamma_{12}c_2 + I_1(t) - \gamma_{21}c_1$$

$$\dot{q}_2 = \gamma_{21}c_1 - \gamma_{12}c_2 - \gamma_{02}c_2 ,$$

where $I_1(t)$ is the input rate of salt. If we convert to mass equations, with constant $f_{ij} \equiv \gamma_{ij}/V_j$, then we get system (2.8) with an input term $[I_1(t), 0]^T$ (T denotes transpose).

SECTION 3.   FIRST-ORDER CHEMICAL REACTIONS

In this section we consider a first-order irreversible chemical reaction and how it is modeled by compartmental equations [161], [169], [241]. A common industrial process is treated: the reduction of limestone into its main products, calcium oxide and magnesium oxide, through the chemical reactions

(3.1)
$$CaCO_3 \xrightarrow{k_1} CaO + CO_2$$
$$MgCO_3 \xrightarrow{k_2} MgO + CO_2$$

with the respective positive rate constants k_i. We assume that the limestone consists of a fraction β of $CaCO_3$ and a remaining fraction $1 - \beta$ of $MgCO_3$ $(0 < \beta < 1)$. Each mole of reactant which decomposes yields one mole of product plus one mole of carbon dioxide; CO_2 does not affect the rates at which the reactions occur and its production is not of interest.

The reactions of (3.1) are carried out in a vessel which is kept at a constant high temperature. Let time t start at zero, and suppose the functions $q_1(t)$, $q_2(t)$, $q_3(t)$, and $q_4(t)$ denote the mass of $CaCO_3$, $MgCO_3$, CaO, and MgO, respectively, in the vessel at any time t. Let the scalar function u(t) be the rate at which limestone is added to the vessel. If we assume the first-order kinetics that the reaction rate is proportional (with proportionality constant k_i) to the mass of reactant, then the equations

(3.2)
$$\dot{q}_1(t) = \beta u(t) - k_1 q_1(t)$$
$$\dot{q}_2(t) = (1 - \beta)u(t) - k_2 q_2(t)$$

describe the reactions of (3.1) via the Law of Mass Action [151] (also see Section 5). In like manner, we assume the formation of the reaction products is governed by the set of equations

(3.3)
$$\dot{q}_3(t) = k_1 q_1(t)$$
$$\dot{q}_4(t) = k_2 q_2(t) .$$

Let $f_{21} \equiv k_1$ and $f_{42} \equiv k_2$. Then together in matrix-vector form, equations (3.2) and

(3.3) give the <u>limestone reduction model</u>:

(3.4)

$$
\begin{bmatrix} \dot{q}_1 \\ \dot{q}_2 \\ \dot{q}_3 \\ \dot{q}_4 \end{bmatrix} = \begin{bmatrix} -f_{31} & 0 & 0 & 0 \\ 0 & -f_{42} & 0 & 0 \\ f_{31} & 0 & 0 & 0 \\ 0 & f_{42} & 0 & 0 \end{bmatrix} \begin{bmatrix} q_1 \\ q_2 \\ q_3 \\ q_4 \end{bmatrix} + \begin{bmatrix} \beta \\ 1-\beta \\ 0 \\ 0 \end{bmatrix} u .
$$

System (3.4) is a compartmental model, and as such has the diagram in Figure 3.1.

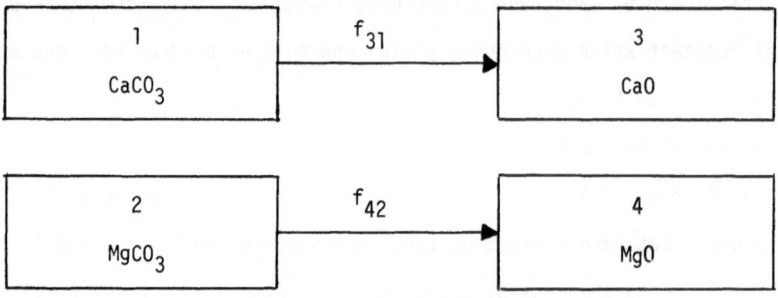

Figure 3.1

In system (3.4), as in the previous systems (2.5) and (2.8), the model has a coefficient matrix which (a) has constant entries; (b) has nonnegative offdiagonal entries; (c) has nonpositive column sums (which requires each diagonal element to be nonpositive). Later we will study general properties of such matrices (e.g., stability) in order to determine how the input function u and the parameters β, f_{31}, and f_{42} influence the solution.

<u>Exercise. 3.1.</u> The simple structure of the matrix in (3.4) allows us to easily analyze the limestone reduction model. Investigate the solution of (3.4) in the case where $u(t) \equiv 0$, but where there is initially M moles of limestone in the reaction vessel. Then consider the situation when $q_i(0) = 0$ for all i, but $u \neq 0$. Why is the system <u>marginally stable</u> [222]? What happens to the solution of (3.4) if limestone is added to the vessel at a constant rate $u(t) \equiv R$ for all $t \geq 0$?

SECTION 4. ENVIRONMENTAL STUDIES

4A. Kinetics of Lead in the Body

In this subsection a compartmental model for the kinetics of lead in the human body is discussed [20]. With lead concentrations in the environment quite high due to industrialization, the effect on human health has been questioned. Some studies suggest that lead is carcinogenic. For instance, a major source of lead is car exhaust, and it has been shown that people who have been greatly exposed to exhaust fumes have a relatively high incidence of cancer.

Here we model lead distribution in the human body by a continuous deterministic dynamic system which is compartmental. Different studies suggest the basic 3-compartment model as shown in Figure 4.1.

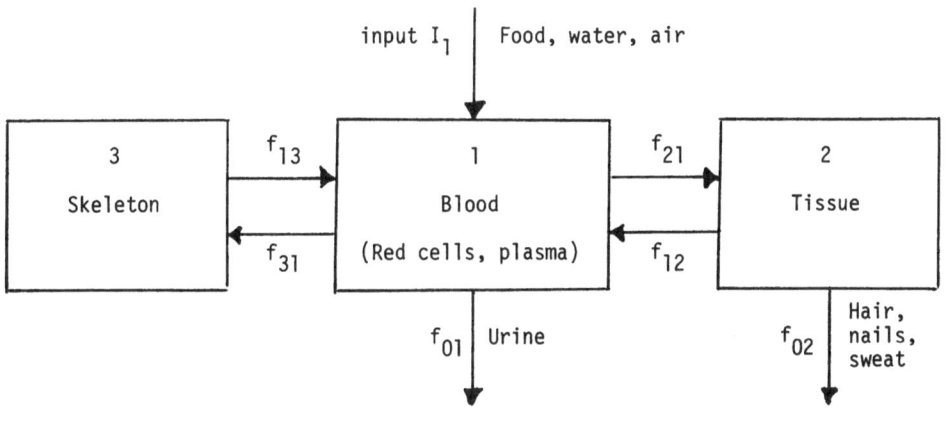

Figure 4.1

Lead enters the body via food and liquid intake as well as by inhalation. From the digestive tract and the lungs there is a large uptake of lead by the red blood cells and to a lesser extent by the plasma. From the blood, lead is fairly rapidly distributed to the tissues (first to the liver and kidney, and later with different time lags to other parts of the body). More slowly, uptake of lead by the bones occurs. This physiology motivates the model of Figure 4.1. Compartments 2 and 3 exchange lead with compartment 1 by diffusion. Thus, as studied in Section 2C, we will assume the diffusion process to be linear (especially for low concentrations of lead). Let $t \geq 0$ be the time variable, $q_i(t)$, $i = 1, 2, 3$, be the amount of lead in compartment i, and $I_1(t)$ be the input rate into compartment 1. Then, by standard arguments, the mass balance equations for the transfer of lead between the bodily compartments are

$$\dot{q}_1 = (I_1 + f_{12}q_2 + f_{13}q_3) - (f_{01}q_1 + f_{21}q_1 + f_{31}q_1)$$

(4.1)
$$\dot{q}_2 = f_{21}q_1 - (f_{02}q_2 + f_{12}q_2)$$

$$\dot{q}_3 = f_{31}q_1 - f_{13}q_3$$

where the f_{ij} are positive constants.

The input rate I_1 in the first equation of (4.1) requires further comment. This rate consists of two components--one from the lungs and one from the digestive tract. Let α be the constant rate of intake of lead into the lungs from the environment. Only a portion p $(0 < p < 1)$ of this input α is actually absorbed into the blood, so that the constant input rate of lead into compartment 1 from the lungs is $p\alpha$. Let β be the constant input rate of lead into the digestive tract via food and water intake. (Also there is some lead intake to the digestive tract from saliva, gastric secretion, and bile, but we do not consider that here.) Of this input β, only a fraction $r\beta$ will be absorbed through the digestive tract to the blood $(0 < r < 1)$. Thus input rate I_1 in (4.1) is the constant

$$I_1 = p\alpha + r\beta .$$

Hence (4.1) can be put into the form

(4.2)
$$\begin{bmatrix} \dot{q}_1 \\ \dot{q}_2 \\ \dot{q}_3 \end{bmatrix} = \begin{bmatrix} f_{11} & f_{12} & f_{13} \\ f_{21} & f_{22} & 0 \\ f_{31} & 0 & f_{33} \end{bmatrix} \begin{bmatrix} q_1 \\ q_2 \\ q_3 \end{bmatrix} + \begin{bmatrix} p\alpha + r\beta \\ 0 \\ 0 \end{bmatrix}$$

where

$$f_{11} \equiv - (f_{01} + f_{21} + f_{31})$$

(4.3) $$f_{22} \equiv - (f_{02} + f_{12})$$

$$f_{33} \equiv - f_{13} .$$

As we shall discuss in considerable detail later, one of the important problems in biological models of this type is the <u>identification of the model</u> and the accompanying <u>parameter estimation</u>. That is, we need to determine the parameters of model (4.2), and also ask whether or not they can be uniquely determined. This is done through a <u>standard tracer experiment</u> carried out on a particular human subject. Before the start of the experiment, the person's body is assumed to be in steady-state with respect to lead, defined by $\dot{q}_{iss} = 0$, $i = 1, 2, 3$, for all t. A set of corresponding values q_{1ss}, q_{2ss}, q_{3ss} is determined beforehand. Then for a prescribed number of days (the unit of time), part of the dietary lead for the subject is replaced by a small amount of a stable lead isotope, e.g., ^{204}Pb (the tracer). This isotope provides an observable transient in the body that can be measured over time. From the measurements gathered, estimates of the parameters in model (4.2) can then be made. This procedure is fully discussed in Section 7 and demonstrated in Section 8.

The following measurements have been made on a particular individual [20] ($\mu g \equiv$ micrograms, $d \equiv$ days):

$$q_{1ss} = 1800 \ (\mu g), \quad q_{2ss} = 700, \quad q_{3ss} = 200000 .$$

The steady-state flows out of compartment 1 are measured as (units $\mu g/d$)

$$f_{01}q_{1ss} = 38, \quad f_{21}q_{1ss} = 20, \quad f_{31}q_{1ss} = 7.$$

Thus we know f_{01}, f_{21}, f_{31}, and also f_{11} from (4.3). In a similar way the parameters f_{ij} for the other compartments are computed.

<u>Exercise 4.1</u>. Suppose the system matrix $F \equiv [f_{ij}]$ of model (4.2) turns out to be

$$
F = \begin{bmatrix} \dfrac{-65}{1800} & \dfrac{1088}{87500} & \dfrac{7}{200000} \\[2.5ex] \dfrac{20}{1800} & \dfrac{-20}{700} & 0 \\[2.5ex] \dfrac{7}{1800} & 0 & \dfrac{-7}{200000} \end{bmatrix}
$$

for the particular subject tested [20]. Why is system (4.2) asymptotically stable [222] in the sense that it is convergent to an equilibrium whenever the input rate I_1 takes on a vew value? Give a measure of the "speed" of this convergence. When air pollution doubles, but dietary intake of lead remains constant, I_1 in this person is found to increase to the value 66.3 (μg/d). Find the new equilibrium vector of system (4.2).

4B. The Aleut Ecosystem

The next example of compartmental modeling deals with a preliminary systems analysis of the Aleut ecosystem in which certain specific questions about that ecological environment are addressed [126]. The Aleutian domain is an island chain in Alaska. The Aleut human inhabitants are considered ecosystem consumers who are at the top of a natural food-chain. In this study carbon is used as a tracer to help elucidate the basic relationships within this complex food-chain. The Aleuts are just one component of this geochemical carbon cycle.

The mathematical model for the system is taken to be $\dot{q} = Fq$ (see tracer kinetics in Section 7), where the dependent variables $q_1(t)$, $q_2(t)$, ..., $q_9(t)$ are the amounts of carbon in

1. the atmosphere

2. land plants

3. dead organic terrestrial matter

4. man

5. phytoplankton

6. zooplankton and marine animals

7. dead organic marine matter

8. surface water

9. the deep sea.

These nine compartments were chosen so as to encompass the total marine carbon cycle of the Aleutian area. The interconnections of the system are shown in Figure 4.2.

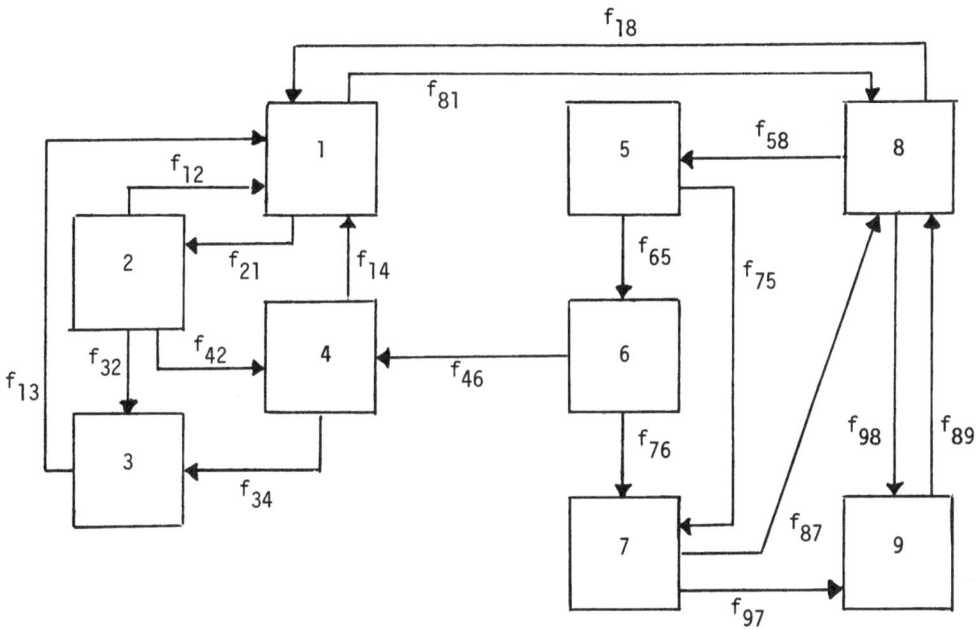

Figure 4.2

The 9-Compartment Aleutian Carbon Cycle

The 9×9 matrix $F \equiv [f_{ij}]$ is identified by using accumulated world figures on amounts of carbon present, and by approximate biomasses. This matrix is (all nonlisted elements are zero)

$$
\begin{bmatrix}
f_{11} & f_{12} & f_{13} & f_{14} & & & & f_{18} & \\
f_{21} & f_{22} & & & & & & & \\
& f_{32} & f_{33} & f_{34} & & & & & \\
& f_{42} & & f_{44} & & f_{46} & & & \\
& & & & f_{55} & & & f_{58} & \\
& & & & f_{65} & f_{66} & & & \\
& & & & f_{75} & f_{76} & f_{77} & & \\
f_{81} & & & & & & f_{87} & f_{88} & f_{89} \\
& & & & & & f_{97} & f_{98} & f_{99}
\end{bmatrix}
$$

with the values

$f_{11} = -1.93 \times 10^{-1}$ $f_{32} = 5.58 \times 10^{-2}$ $f_{58} = 8.00 \times 10^{-2}$

$f_{12} = 2.23 \times 10^{-2}$ $f_{33} = -3.57 \times 10^{-2}$ $f_{65} = 4.00$

$f_{13} = 3.57 \times 10^{-2}$ $f_{34} = 4.58 \times 10^{-1}$ $f_{66} = -6.67$

$f_{14} = 3.32$ $f_{42} = 4.21 \times 10^{-8}$ $f_{75} = 4.00$

$f_{18} = 1.94 \times 10^{-1}$ $f_{44} = -3.78$ $f_{76} = 6.67$

$f_{21} = 5.00 \times 10^{-2}$ $f_{46} = 5.08 \times 10^{-4}$ $f_{77} = -1.30 \times 10^{-2}$

$f_{22} = -7.81 \times 10^{-2}$ $f_{55} = -8.00$ $f_{81} = 1.43 \times 10^{-1}$

$f_{87} = 1.13 \times 10^{-2}$ $f_{88} = -3.54 \times 10^{-1}$ $f_{89} = 1.30 \times 10^{-3}$

$f_{97} = 1.67 \times 10^{-3}$ $f_{98} = 8.00 \times 10^{-2}$ $f_{99} = -1.30 \times 10^{-3}$

Note that the matrix is fairly sparse and the nonzero entries are concentrated about the main diagonal.

Certain questions about the nature of the Aleuts interaction with their environment have been posed:

1. How dependent are the Aleuts on various components of the ecosystem?

2. Are there properties inherent in the ecosystem which make it advantageous for the Aleuts to utilize the marine environment more extensively than the terrestrial environment?

3. How important are the Aleuts in affecting stability of the system?

Using the above model, partial answers to these questions and others were found and are discussed in [126].

SECTION 5.   NONLINEAR COMPARTMENTAL MODELS

5A.   Continuous Flow Chemical Reactor

In this section we look at some nonlinear compartmental models, the first of
which is a continuous flow chemical reactor [161], [134]. This reactor is a device
used in the chemical industry to allow chemicals to react in a controlled environment. A tank of constant volume, which is the reactor vessel, is kept at a constant temperature. There is continuous flow of reactants into this constantly
stirred tank. After certain reactions take place inside the tank, a portion of the
uniform mixture is then drained from the tank. This process is illustrated in
Figure 5.1, where R denotes the fixed rate constant for flow into and out of the
tank.

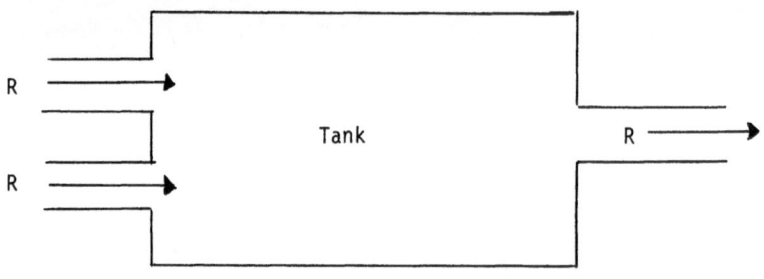

Figure 5.1

For this particular example, suppose that one mole of H_2O and one mole of SO_3
undergo an irreversible chemical reaction in the tank to form one mole of sulfuric
acid H_2SO_4:

(5.1 $H_2O + SO_3 \xrightarrow{k} H_2SO_4$.

Suppose a molar concentration u_1 of H_2O and a molar concentration u_2 of SO_3 are
contained in the upper and lower inputs to the reactor, respectively. Let c_1, c_2,
c_3, denote the molar concentrations of H_2O, SO_3, H_2SO_4, respectively, in the tank.

To quantitatively model the chemical reaction, we use the Law of Mass Action.
We have here the situation wherein a molecule α of one chemical and a molecule β of
a second chemical combine to give a molecule γ of a third chemical. For α and β to

react, they must collide. Not every collision will result in a binding of the molecules, but a plausible assumption is that the production of molecules γ is proportional to the number of encounters of molecules α and β. Thus in terms of the concentrations [·] of the chemicals, we are lead to the equation

$$\frac{d[\gamma]}{dt} = k[\alpha][\beta]$$

representing part of the chemical reaction $\alpha + \beta \xrightarrow{k} \gamma$, with rate constant $k > 0$. We now utilize this in the modeling of the concentration of H_2SO_4 in the reactor by saying that

(5.2) $$\frac{dc_3}{dt} = \text{gain} - \text{loss} = kc_1c_2 - Rc_3.$$

If for the moment we think of the concentration $[\beta]$ as fixed, then it is also plausible to suppose that the frequency of collisions which result in an actual binding with chemical α is proportional to the concentration $[\alpha]$. Hence we would expect

$$\frac{d[\alpha]}{dt} = - k[\alpha][\beta]$$

and by symmetry between α and β,

$$\frac{d[\beta]}{dt} = - k[\beta][\alpha] .$$

The validity of these ideas has been established experimentally [151, p. 303]. Thus for the reactor model, we could say

(5.3) $$\frac{dc_1}{dt} = \text{gain} - \text{loss} = Ru_1 - (Rc_1 + kc_1c_2)$$

and

(5.4) $$\frac{dc_2}{dt} = \text{gain} - \text{loss} = Ru_2 - (Rc_2 + kc_1c_2) .$$

In terms of the amounts $q_i(t) = Vc_i(t)$, $i = 1, 2, 3$, equations (5.2) - (5.4) become

$$\dot{q}_1 = RVu_1 - (Rq_1 + \frac{k}{V} q_1 q_2)$$

(5.5) $$\dot{q}_2 = RVu_2 - (Rq_2 + \frac{k}{V} q_1 q_2)$$

$$\dot{q}_3 = \frac{k}{V} q_1 q_2 - Rq_3 \ .$$

Let

$$f_{11} \equiv - (R + \frac{k}{V} q_2) \qquad\qquad f_{22} \equiv - (R + \frac{k}{V} q_1)$$

$$f_{32} \equiv \frac{k}{V} q_1 \qquad\qquad f_{33} \equiv - R \ .$$

Then equations (5.5) can be put into the compartmental system format

(5.6)
$$\begin{bmatrix} \dot{q}_1 \\ \dot{q}_2 \\ \dot{q}_3 \end{bmatrix} = \begin{bmatrix} f_{11} & 0 & 0 \\ 0 & f_{22} & 0 \\ 0 & f_{32} & f_{33} \end{bmatrix} \begin{bmatrix} q_1 \\ q_2 \\ q_3 \end{bmatrix} + \begin{bmatrix} RVu_1 \\ RVu_2 \\ 0 \end{bmatrix}$$

which corresponds to the compartmental diagram of Figure 5.2 (arrows indicating inputs from the outside environment are labeled with actual input rates RVu_i).

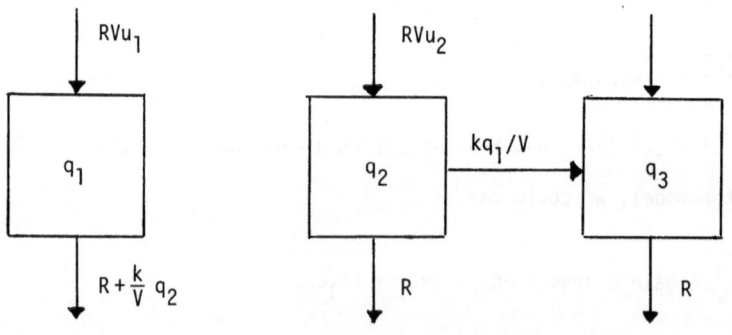

Figure 5.2

System (5.6) is <u>nonlinear</u> since some of the transfer coefficients f_{ij} are functions of certain variables q_k. However, note that the coefficient matrix of

(5.6) still satisfies the basic conditions that have been demonstrated in the models of previous sections: each offdiagonal element is nonnegative, and every column sum is nonpositive.

Exercise 5.1. A particular bimolecular reaction

$$A + B \underset{k_2}{\overset{k_1}{\rightleftarrows}} AB$$

can be modeled by the nonlinear system [134, p. 13]

$$(5.7) \quad \begin{bmatrix} \dot{q}_1 \\ \dot{q}_2 \\ \dot{q}_3 \end{bmatrix} = \begin{bmatrix} -\left(\frac{v}{V} + \frac{k_1}{V} q_2\right) & 0 & k_2 \\ 0 & -\left(\frac{v}{V} + \frac{k_1}{V} q_1\right) & k_2 \\ \frac{k_1}{V} q_2 & 0 & -\left(k_2 + \frac{v}{V}\right) \end{bmatrix} \begin{bmatrix} q_1 \\ q_2 \\ q_3 \end{bmatrix} + \begin{bmatrix} v_1 c_1 \\ v_2 c_2 \\ 0 \end{bmatrix}$$

where q_i and c_i are the only functions of time. What must be done to the system matrix of (5.7) so that it satisfies the conditions stated in the last paragraph? [Hint: Consider adding a fictitious input to the third equation of (5.7).]

5B. Reaction Order

Each term of equations (5.5) (other than input terms) describes what is called a unit process. For any such unit process, e.g., term $kq_1(t)q_2(t)$, we sum the exponents of all the dependent variables in that term to get the order of the process. For instance, terms of the form $kq_1 q_2$ in (5.5) each describe a second-order process. Most particle transitions are of the orders 0, 1, or 2. Higher ordered transitions are more rare.

If the time rate of change of the number of particles $q(t)$ is completely independent of the number of particles present in a particular compartment, then $dq/dt =$ constant $\equiv k$, and the transition process is zeroeth-order. The rate dq/dt is instead determined by other factors, such as the amount of catalyst present. When the transition of a type of particle is not affected by other particles, it undergoes a first-order process: $dq/dt = kq$. If two particles bind in a transition, then the process is governed by $dq_i/dt = kq_j q_\ell$ and is second-order. Reactions which require the interaction or collision of two molecules to produce a single product molecule are often in this category (as we have seen in Section 5A).

Exercise 5.2.   Show that the solution of the second-order model

$$dq/dt = kq^2, \qquad t \geq 0 ,$$

is

(5.8) $q(t) = q(0)/[1 - q(0)kt]$.

In many studies detailed knowledge about the mechanism of the process, that is, the order of the process, is often important to know. How does one determine that order? As an example, suppose equation (5.8) is considered in the form

(5.9) $k = \dfrac{1}{t} \left[\dfrac{1}{q(0)} - \dfrac{1}{q(t)} \right]$, $t > 0$,

and we have run an experiment in which the data points in the first two rows in Table 5.1 have been collected. Now if the process order is two, then all data points $\{(t_i, q(t_i))\}$ should satisfy equation (5.9) for some constant k. Hence we should get more or less the same values for $k_1, k_2, \ldots, k_6$.

TABLE 5.1

t	0	t_1	t_2	t_3	t_4	t_5	t_6
q(t)	q(0)	$q(t_1)$	$q(t_2)$	$q(t_3)$	$q(t_4)$	$q(t_5)$	$q(t_6)$
k from (5.9)		k_1	k_2	k_3	k_4	k_5	k_6

5C. Other Nonlinear Compartmental Models

    In the mathematical theory of epidemics, a simple deterministic model for the spread of an epidemic is the nonlinear compartmental model

$$(5.10) \qquad \begin{bmatrix} \dot{q}_1 \\ \\ \dot{q}_2 \end{bmatrix} = \begin{bmatrix} f_{11} & 0 \\ \\ f_{21} & f_{22} \end{bmatrix} \begin{bmatrix} q_1 \\ \\ q_2 \end{bmatrix} + \begin{bmatrix} -u \\ \\ 0 \end{bmatrix}$$

where $f_{11} \equiv -\beta q_2$, $f_{21} \equiv \beta q_2$, $f_{22} \equiv -\gamma$, and β, γ are positive parameters [161], [155]. Here there are two groups of individuals in a fixed population: susceptibles and infectives. Let $q_1(t)$ be the number of susceptibles, and let $q_2(t)$ be the number of infectives, each at time $t \geq 0$. The disease is spread by direct contact between individuals of the two classes; a plausible assumption is that the rate at which susceptibles take the disease and become infective is $\beta q_1 q_2$ for some constant β. This is analogous to the argument presented earlier in Section 5A for the Law of Mass Action. Infectives can die or become immune to the disease at a rate $-\gamma q_2$ for some constant γ. Moreover, susceptibles may be reduced in the population at a rate $-u(t)$ due to the introduction of vaccinations. Thus we arrive at equations (5.10).

Nonlinear compartmental models arise in <u>enzyme kinetics</u> [151]. Enzymes are large protein molecules that can catalyze chemical reactions. The enzyme molecule e combines with the input or substrate molecule s to form a molecule or complex c. When bound to the enzyme molecule in this complex, the substrate is somehow more likely to form the output or product p of the reaction. When this occurs the complex breaks into the product molecule and the enzyme molecule. It is presumed to follow the reaction schematic

$$s + e \underset{k_{-1}}{\overset{k_1}{\rightleftharpoons}} c \overset{k_2}{\longrightarrow} p + e \,.$$

Standard arguments, as we have discussed before in Section 5A, lead to the nonlinear compartmental model [151, p. 305]

$$\begin{bmatrix} \dot{s} \\ \\ \dot{c} \end{bmatrix} = \begin{bmatrix} f_{11} & f_{12} \\ \\ f_{21} & f_{22} \end{bmatrix} \begin{bmatrix} s \\ \\ c \end{bmatrix}$$

where $f_{11} \equiv - k_1\bar{e}$, $f_{12} \equiv k_1s + k_{-1}$, $f_{21} \equiv k_1\bar{e}$, $f_{22} \equiv - (k_1s + k_{-1} + k_2)$, and $\bar{e}$ is $e(t) + c(t)$, which turns out to be constant over all time t.

A related nonlinear 3-compartment model is found in modeling the kinetics of the enzyme cytochrome P-450 [90] (also see the recently published article: K. Tuttle, J. A. Peterson, L. Peterson, J. Eisenfeld, Identification of nonlinear compartmental systems with an application to the modelling of the enzyme cytochrome P-450, in Nonlinear Phenomena in Mathematical Sciences, V. Lakshmikantham (ed.), Academic Press, New York, 1982, 351-362). The closed system is described by the model $\dot{q}$ = Fq where

$$F \equiv [f_{ij}] \equiv \begin{bmatrix} -a_{21}(q_1 + \sigma) & a_{12} & 0 \\ a_{21}(q_1 + \sigma) & -(a_{12} + a_{32}) & a_{23} \\ 0 & a_{32} & -a_{23} \end{bmatrix} ,$$

q(t) is the vector $[q_1(t), \quad q_2(t), \quad q_3(t)]^T$, and $\sigma > 0$ is a parameter.

As another example of a nonlinear compartmental system, we refer the reader to [2] wherein the set

$$A_1 + A_2 \xrightarrow{k_1} A_3$$

$$A_3 + A_2 \xrightarrow{k_2} A_4$$

$$A_5 + A_2 \xrightarrow{k_3} A_6$$

$$A_6 + A_2 \xrightarrow{k_4} A_7$$

of chemical reactions is modeled. The reactions proceed in a setting of constant volume and temperature. It is assumed (as before) that the rate of change of the concentration of each substance in any one reaction is proportional to the product of the concentrations of substances being used up in the reaction. If k_i, i = 1, 2, 3, 4, are proportional to the respective rate constants, and $q_i(t)$ are the amounts of the chemicals, then the equations of the compartmental model can be written in matrix-vector form as $\dot{q} = Fq$, where

$$F \equiv \begin{bmatrix} f_{11} & 0 & 0 & 0 & 0 & 0 & 0 \\ 0 & f_{22} & 0 & 0 & 0 & 0 & 0 \\ f_{31} & 0 & f_{33} & 0 & 0 & 0 & 0 \\ 0 & 0 & f_{43} & 0 & 0 & 0 & 0 \\ 0 & 0 & 0 & 0 & f_{55} & 0 & 0 \\ 0 & 0 & 0 & 0 & f_{65} & f_{66} & 0 \\ 0 & 0 & 0 & 0 & 0 & f_{76} & 0 \end{bmatrix}$$

and

$$f_{11} \equiv - k_1 q_2$$

$$f_{22} \equiv - (k_1 q_1 + k_2 q_3 + k_3 q_5 + k_4 q_6)$$

$$f_{31} \equiv k_1 q_2$$

$$f_{33} \equiv - k_2 q_2$$

$$f_{43} \equiv k_2 q_2$$

$$f_{55} \equiv - k_3 q_2$$

$$f_{65} \equiv k_3 q_2$$

$$f_{66} \equiv - k_4 q_2$$

$$f_{76} \equiv k_4 q_2 \; .$$

One final note. It is known that the types of nonlinear equations presented in this section, $\dot{q} = F(q)q + I$, along with the appropriate initial condition $q(0)$ at $t = 0$, do indeed have <u>unique</u> solutions which are nonnegative whenever $q(0) \geq 0$ [23].

SECTION 6. THE GENERAL COMPARTMENTAL MODEL

Suppose we have a compartmental system consisting of compartments numbered from 1 to n. The general description for the dynamics of the exchange of material with respect to the ith compartment is the mass balance equation

(6.1) $\dot{q}_i$ = rate of inflow - rate of outflow   i = 1, 2, ..., n,

where $q_i(t) \geq 0$ is the quantity of material in compartment i at time t. The rate of transfer of material from compartment j to compartment i (i ≠ j) is modeled by $f_{ij}q_j$, where f_{ij} is a nonnegative quantity called a fractional transfer coefficient which may be a function of the amounts $q \equiv [q_1, q_2, ..., q_n]^T$ (T denotes transpose), time t, and a vector of parameters $\alpha = [\alpha_1, \alpha_2, ..., \alpha_\nu]^T$. Hence, from (6.1), the general compartmental model is

(6.2)

$$\dot{q}_i(t) = \sum_{\substack{j=1 \\ j \neq i}}^{n} f_{ij}(q(t), t, \alpha)q_j(t) + I_i(t)$$

$$- \sum_{\substack{j=0 \\ j \neq i}}^{n} f_{ji}(q(t), t, \alpha)q_i(t)$$

over some time interval $[0, t_1]$. The function $I_i(t)$ is the rate of input of material into the ith compartment from the outside, and f_{0i} is the fractional excretion coefficient so that $f_{0i}q_i$ is the rate of excretion of material to the outside environment from the ith compartment. A block diagram for a two compartment system illustrating exchanges of material is given in Figure 6.1.

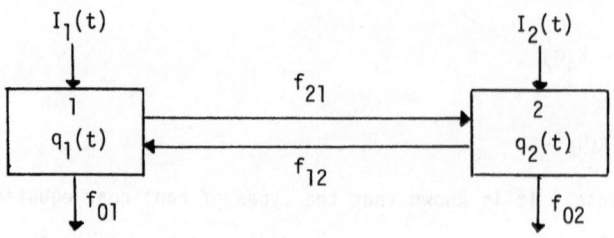

Figure 6.1

It is convenient to lump together all coefficients in the second summation of (6.2) and define

$$(6.3) \qquad f_{ii} \equiv - \sum_{\substack{j=0 \\ j \neq i}}^{n} f_{ji} \qquad\qquad i = 1, 2, \ldots, n.$$

The total outflow from compartment i to other compartments and the outside environ-
ment is $f_{ii}q_i$. The elements f_{ij} ($i \neq j$) are nonnegative since they correspond to inflows while the f_{ii} are nonpositive since they measure outflows. In matrix-
vector form, the general compartmental model (6.2) can be written as the ordinary differential equation system

$$(6.4) \qquad \dot{q} = Fq + I$$

in which $F \equiv [f_{ij}]$ is the $n \times n$ matrix of fractional transfer coefficients, and $I \equiv [I_1, I_2, \ldots, I_n]^T$.

In system (6.2) the fractional transfer coefficients meet certain constraints imposed by the biological model. Quite often we encounter compartmental systems for which all f_{ij} are <u>constants</u> (which depend only on certain parameters of the model). These have been illustrated in some previous sections, and we also run up against them in tracer experiments, which will be discussed in the next section. The analysis of the linear system form of (6.2) with constant coefficients is the major theme of this monograph. In other instances f_{ij} is a function of q_i, q_j and some parameters, but not directly dependent on time so that the compartmental model (6.2) is <u>autonomous</u>. When the f_{ij} are <u>time dependent</u>, it usually takes the form of periodic coefficients [134, p. 145], or the coefficients for the kinetics of distribution of a tracer in linear systems which are not in steady-state [134, p. 74].
If the exchange of material from compartment j to compartment i takes some positive amount of time σ_{ij}, this can be built into the model as a <u>time delay</u>, and the first summation in (6.2) is replaced by

$$\sum_{\substack{j=1 \\ j \neq i}}^{n} f_{ij} \; \dot{q}_j(t-\sigma_{ij}).$$

Exact solution of the nonlinear equations (6.2) is either quite complex or impossible. However, we are usually not interested in the complete solution, only the steady-state solution. This is defined as the solution of the differential equations as $t \to \infty$. If the inputs I_i are constant, then the steady-state is associated with nonchanging quantities and so $dq_i/dt = 0$ for all i.

The compartmental models (6.2) found in applications are normally of a special type [134], [135]. The f_{ij} are functions of q only and dependent on the α-parameters. As a function of each individual q_k, f_{ij} is bounded and is either increasing to a positive limit, or decreasing to a nonnegative limit, as q_k increases. Compartmental models of this type, as restricted classes of general nonlinear ordinary differential equation systems, need further analysis and study, and are of special interest to bioscientists.

SECTION 7. TRACER KINETICS IN STEADY STATE SYSTEMS

7A. The Tracer Equations

In this section, we describe a <u>standard experimental technique</u> involving a
tracer study of a compartmental system. The method is to bring the system, linear
or nonlinear, to a steady-state, inject a trace compound (in a negligible amount
compared to the unlabeled species), and follow its distribution throughout the system. As we will see, the tracer distribution follows <u>linear</u> kinetics.

The exchange of unlabeled material is governed by the general compartmental
equations

$$(7.1a) \qquad \dot{q}_i(t) = \sum_{j=1}^{n} f_{ij}(q(t), t, \alpha)q_j(t) + I_i(t) \qquad\qquad i = 1, 2, \ldots, n.$$

For ease we deal with a two compartment system where we suppose that the inputs I_1
and I_2 are constants, and the above equations are represented in the form

$$(7.1b) \qquad \dot{q}_i = F_i(q_1, q_2) + I_i \qquad\qquad i = 1, 2.$$

If q_{iss} is the constant amount of unlabeled material in compartment i under this
steady-state condition, then that quantity satisfies the equations

$$(7.2) \qquad 0 = \dot{q}_{iss} = F_i(q_{1ss}, q_{2ss}) + I_i \qquad\qquad i = 1, 2.$$

Now suppose, at time zero, that the system is pushed off the steady-state by
the sudden injection of a small amount of tracer which is distinguishable from the
tracee. After a while, the system returns to the steady-state. This input of
tracer may be in the form of a bolus input instantaneous at $t = 0$, or the tracer may
be continuously infused for a short period of time. Let the input rate of tracer
into compartment i at time $t \geq 0$ be denoted by $b_i(t)$. Because each compartment is
<u>well-stirred</u>, once tracer enters a compartment it is assumed to be <u>immediately</u>
spread throughout the compartment so that at all times there is a homogeneous mixture of tracer and tracee. We note that there are three major contributors to rapid
distribution of the tracer in a compartment: (a) stirring or <u>mixing</u> by currents

within the body of tracee; (b) transportation (convection) by the flowing stream of unlabeled material; (c) diffusion or molecular motion within the mixture of tracer and tracee. Under these assumptions, we can show that the differential equations which describe the distribution of the tracer in the system are linear with constant coefficients.

The radioisotopes used as tracers are different in mass than their normal isotopes. If the mass differences are small, then in the exchange of material between compartments there is little discrimination favoring the tracer over the tracee, or vice versa. This negligible isotope fractionation effect is necessary to get accurate estimates of the fractional transfer coefficients [134, p. 72].

Let the amount of tracer in compartment i at any time $t \geq 0$ be denoted by $x_i(t)$. We assume that the quantity of tracer injected is of sufficiently small volume so as to not disturb the steady-state characteristics of the system. Since tracer and tracee are indistinguishable, except for the fact that one is labeled, we assume that the total quantity Q_i of tracer and tracee in the ith compartment follows the equations

$$(7.3) \quad \begin{aligned} \dot{Q}_1 &= F_1(Q_1, Q_2) + I_1 + b_1 \\ \dot{Q}_2 &= F_2(Q_1, Q_2) + I_2 + b_2 \end{aligned}$$

in keeping with equations (7.1) and no isotope fractionation effect. Let us scale the amounts of tracer,

$$(7.4) \qquad x_i(t) = Q_i(t) - q_{iss} \qquad i = 1, 2,$$

by a small real number ε $(0 < |\varepsilon| \ll 1)$; introduce $y_i(t)$ defined by

$$(7.5) \qquad x_i(t) \equiv \varepsilon y_i(t) \qquad (i = 1, 2).$$

Then equations (7.3), upon the substitution of (7.4) and (7.5), become $(i = 1, 2)$

$$(7.6) \qquad \varepsilon \dot{y}_i = F_i(q_{1ss} + \varepsilon y_1, q_{2ss} + \varepsilon y_2) + I_i + b_i .$$

We now use the Taylor series expansion for a function $F(x,y)$ of two variables:

$$F(x_0 + h, y_0 + k) = F(x_0, y_0) + h \frac{\partial F}{\partial x}(x_0, y_0) + k \frac{\partial F}{\partial y}(x_0, y_0) + \ldots .$$

In this manner, (7.6) becomes (i = 1, 2)

(7.7)
$$\varepsilon \dot{y}_i = F_i(q_{1ss}, q_{2ss}) + \varepsilon y_1 \frac{\partial F_i}{\partial q_1}(q_{1ss}, q_{2ss})$$

$$+ \varepsilon y_2 \frac{\partial F_i}{\partial q_2}(q_{1ss}, q_{2ss}) + O(\varepsilon^2) + I_i + b_i ,$$

where the symbol $O(\varepsilon^2)$ is read as "order ε^2" and indicates that the order of magni-

tude of the most important neglected term is ε^2. Recalling that the steady state

condition (7.2) is in force, the following set of equations is obtained from (7.7)

by neglecting $O(\varepsilon^2)$ terms (since the amount of tracer is small):

(7.8) $\qquad \dot{x}_i = \frac{\partial F_i}{\partial q_1}(q_{1ss}, q_{2ss}) x_1 + \frac{\partial F_i}{\partial q_2}(q_{1ss}, q_{2ss}) x_2 + b_i$

for i = 1, 2. This is a linear system of ordinary differential equations in x_1 and

x_2 with constant coefficients

(7.9) $\qquad a_{ij} \equiv \frac{\partial F_i}{\partial q_j}(q_{1ss}, q_{2ss}) \qquad (i,j = 1, 2).$

Thus we have shown that the <u>essential qualitative features</u> of equations (7.1) under

steady-state conditions are given by the solution of the linearized constant coef-

ficient version (7.8). As a consequence, the <u>mass balance equations for the tracer</u>

are (t ≥ 0)

(7.10)
$$\dot{x}_1 = a_{11}x_1 + a_{12}x_2 + b_1$$

$$\dot{x}_2 = a_{21}x_1 + a_{22}x_2 + b_2$$

along with the appropriate initial conditions. The constants a_{ij} reflect the steady-

state characteristics of the linear or nonlinear compartmental system since, by

(7.9), they are nonlinear algebraic functions of the steady state tracee values. If

there are no isotope fractionation effects, then within first order approximation, we have

(7.11) $f_{ij} = a_{ij}$ $i,j = 1, 2.$

This result is quite important, for <u>through a tracer experiment (7.11) provides a</u> <u>method of estimating the steady-state fractional transfer coefficients f_{ij} of the</u> <u>compartmental model</u> [134], [85].

Exercise 7.1.   Carry out the development given above in going from (7.1) to (7.10) for the particular set of equations

$$\dot{q}_1 = - k_1 q_1 + I_1$$

$$\dot{q}_2 = - k_2 q_2^2 + k_3 q_1 + I_2$$

with constants k_i and I_i positive for all i.   Show that the coefficients in (7.9) or (7.10) are

$$a_{11} = - k_1 \qquad\qquad a_{12} = 0$$

$$a_{21} = k_3 \qquad\qquad a_{22} = - 2k_2 q_{2ss} \,.$$

7B. Linear Compartmental Models

The analysis of tracer kinetics given above in the two variable case immediately generalizes to a system with n compartments. The kinetics of the tracer amount $x_i(t)$ will be assumed to follow the constant coefficient linear equation

$$(7.12) \qquad \dot{x}_i(t) = b_i(t) + \sum_{\substack{j=1 \\ j \neq i}}^{n} a_{ij} x_j - \sum_{\substack{j=0 \\ j \neq i}}^{n} a_{ji} x_i$$

for each i = 1, 2, ..., n. These equations are just special cases of form (7.1a). The $b_i(t)$ is the input rate of tracer to compartment i from outside the system. The constant fractional transfer coefficients a_{ij} (i ≠ j) are nonnegative. The last sum in (7.12) can be rearranged as

$$\left(- \sum_{\substack{j=0 \\ j \neq i}}^{n} a_{ji} \right) x_i \, ,$$

which leads to the nonpositive constant

(7.13) $$a_{ii} \equiv - \sum_{\substack{j=0 \\ j \neq i}}^{n} a_{ji}$$

for each $i = 1, 2, \ldots, n$. Equation (7.13) fills out the remaining entries of an $n \times n$ constant matrix $A \equiv [a_{ij}]$. This matrix has the properties that

(a) every off-diagonal entry is nonnegative;

(b) each diagonal entry is nonpositive;

(c) the sum of any column, say the jth column, is the nonpositive number
     $- a_{0j}$ (since (7.13) holds).

Any square matrix consisting of constant elements possessing properties (a)-(c) will be known as a <u>compartmental matrix</u>. A simple 4-compartment example is

$$A = \begin{bmatrix} -0.2 & 0.5 & 0.0 & 0.2 \\ 0.0 & -0.9 & 0.0 & 0.2 \\ 0.1 & 0.3 & 0.0 & 0.4 \\ 0.1 & 0.0 & 0.0 & -0.8 \end{bmatrix}.$$

which corresponds to Figure 7.1. Here the second column sum is the negative of $a_{02} = 0.1$. All other column sums (and thus the a_{0j} s) are zero. Because of (7.13), we have

$$|a_{ii}| \geq \sum_{\substack{j=1 \\ j \neq i}}^{n} |a_{ji}| \qquad\qquad i = 1, 2, \ldots, n,$$

so that A is a (column) <u>diagonally dominant matrix</u> for each n.

   Let A be an $n \times n$ compartmental matrix. If we define

$$b(t) \equiv [b_1(t), b_2(t), \ldots, b_n(t)]^T \, ,$$

$$x(t) \equiv [x_1(t),\ x_2(t),\ \ldots,\ x_n(t)]^T ,$$

$$\dot{x}(t) \equiv [\dot{x}_1(t),\ \dot{x}_2(t),\ \ldots,\ \dot{x}_n(t)]^T ,$$

then the following continuous deterministic model governs the tracer kinetics in a general n-compartmental system in steady-state:

(7.14)

$$\dot{x}(t) = Ax(t) + b(t),\ t \geq 0$$

$$x(0) = x_0 .$$

In the terminology of modern dynamic systems theory, the amounts of tracer, $x_i(t)$, i = 1, 2, ..., n, are referred to as the state variables of the system, the matrix A is called the system matrix, and the input vector b(t) is known as the forcing function. The primary purpose of this monograph is to study model (7.14) when the system matrix A is compartmental.

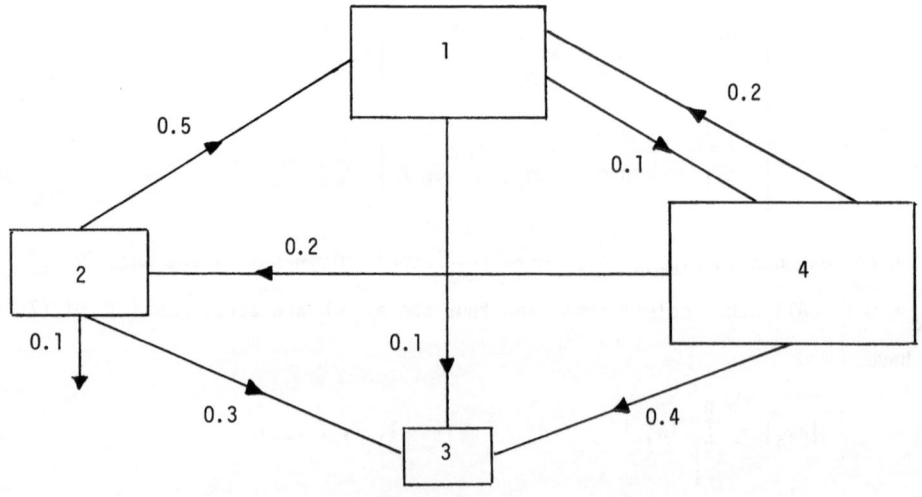

Figure 7.1

Exchange of Tracer for the 4-Compartment Example.

No Flow Goes Out of the 3rd Compartment Since
the Third Column of A Has Only Zero Elements

SECTION 8.   UPTAKE OF POTASSIUM BY RED BLOOD CELLS

In the human bloodstream, potassium ions are continually moving from the
plasma into the red blood cells and vice versa.   The determination of the rates of
potassium influx and efflux is of interest to biomedical scientists in their attempt
to fully comprehend the function of the red blood cells, and to carry on the fight
against blood diseases [212].

The physiology of the bloodstream suggests two compartment behavior of the
system.   It is known that potassium is concentrated in red blood cells by an active
transport mechanism which is nonlinear.   Thus a nonlinear compartmental system is to
be anticipated.   Experimental data reported by Sheppard and Martin [212] show that
potassium levels in both the plasma and red cells stay relatively constant over
time.   Hence we can assume a system in steady-state; so for a tracer the differential equations which describe its distribution are linear.   The totality of red
blood cells containing an amount q_1 of potassium ions (the <u>tracee</u>) will be designated as the first compartment.   Blood plasma, which contains a quantity q_2 of
potassium, is the second compartment.   The flux between these compartments is shown
schematically in Figure 8.1.

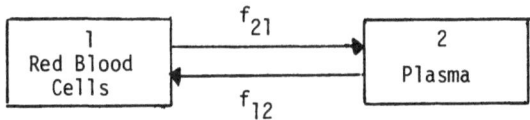

Figure 8.1

Closed Two Compartment System Representing
the Transfer of Potassium Ions

The kinetics of the steady-state system for the flow of potassium ions between the
two compartments is governed by the mathematical model

$$\dot{q}_1 = -f_{21}(q,t)q_1 + f_{12}(q,t)q_2$$

(8.1)

$$\dot{q}_2 = f_{21}(q,t)q_1 - f_{12}(q,t)q_2$$

where $q \equiv [q_1 \quad q_2]^T$. Now by the steady-state assumption, q_1, q_2, f_{21} and f_{12} are all <u>constant</u>.

At time zero, a small known dose D of radioactive K^{42+} ions (the <u>tracer</u>) is injected into the blood plasma. Subsequently the tracer moves back and forth through the red blood cell surfaces, following the movement of the tracee. Let $x_i(t)$ denote the amount of radioactive potassium ions present in compartment i at time $t \geq 0$. At certain postzero times t_1, t_2, ..., t_m, blood plasma samples are drawn and the amounts $\hat{x}_2(t_1)$, $\hat{x}_2(t_2)$, ..., $\hat{x}_2(t_m)$ of radioactivity in each of these samples is measured. These are represented in Figure 8.2. From this data, we hope to determine the permeability (i.e., f_{12} and f_{21}) of the red blood cell surfaces to potassium.

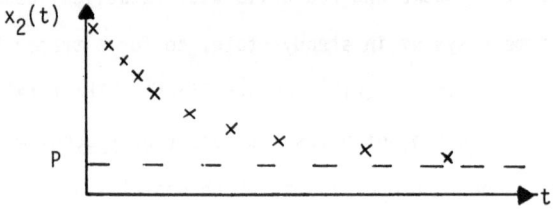

Figure 8.2

Uptake of K^{42+} by Red Blood Cells From Plasma

<u>Exercise 8.1</u>. Show that the kinetics of the <u>tracer</u> is modeled by the equations

(8.2)
$$\dot{x}_i = a_{i1}x_1 + a_{i2}x_2 \qquad (i = 1, 2)$$

$$x_1(0) = 0, \quad x_2(0) = D$$

where $a_{11} = -f_{21}$, $a_{12} = f_{12}$, $a_{21} = f_{21}$, and $a_{22} = -f_{12}$.

With the sampling of just the second compartment, we will find that system (8.2) is identifiable as to the a_{ij}-values. To get the unique estimates of a_{21} and a_{12}, the following approach can be taken. A basic assumption in this model is that the total amount of tracer in the system remains constant (since the system is closed and therefore there can be no loss of material). Hence $x_1(t) + x_2(t) = D$ for all $t \geq 0$. With the aid of the relation $x_1 = D - x_2$, the second equation in (8.2) is rewritten as

$$(8.3) \qquad \dot{x}_2 = f_{21}(D - x_2) - f_{12}x_2 .$$

By this maneuver, the unknown x_1 has been eliminated from the problem. Equation (8.3) involves only the function x_2, which has been observed at discrete times, and the unknown constants f_{21}, f_{12}. Using the initial condition $x_2(0) = D$, the solution of (8.3) is

$$(8.4) \qquad x_2(t) = P[1 + (f_{12}/f_{21}) \exp(mt)],$$

where the parameters

$$(8.5) \qquad P \equiv f_{21}D/(f_{21} + f_{12}) \qquad\qquad m \equiv - (f_{21} + f_{12}) .$$

Exercise 8.2.   Verify the solution (8.4).

In order to determine the rate constants f_{12} and f_{21}, rewrite (8.4) as

$$(8.6) \qquad \ell n[(x_2(t)/P) - 1] = mt + b$$

where

$$(8.7) \qquad b \equiv \ell n(f_{12}/f_{21}) .$$

Denote the left side of (8.6) by the function $f(t)$, $t \geq 0$. We have rewritten equation (8.4) in the form (8.6) because the function f has a straight line graph. Let $\hat{f}(t_i)$ be the values of f when the observed quantities $\hat{x}_2(t_i)$ replace $x_2(t)$, $i = 1, 2, \ldots, m$.

Exercise 8.3.   Consider the data in Table 8.1. Does $f(t)$ have roughly a straight line graph for this data? An important point here is that the model is actually validated provided the data points come close to lying along a straight line.

TABLE 8.1

t	0	250	700	1200	1700
$x_2/P-1$	5	3.2	1.6	0.6	0.2

Exercise 8.4. Show that the value P, presently unknown, is the asymptotic amount of tracer in the second compartment as $t \to \infty$. It is possible to determine P experimentally (see Figure 8.2).

Exercise 8.5. Fit a straight line $y = mt + b$ through the pairs of points given in Table 8.1 to obtain estimates $\hat{m}$ and $\hat{b}$ for m and b, respectively.

Once estimates $\hat{m}$ and $\hat{b}$ are found, then unique values for f_{21} and f_{12} are readily obtainable. From (8.5) and (8.7) we get the nonlinear equations

$$f_{21} + f_{12} = -\hat{m} \qquad f_{12}/f_{21} = \exp(\hat{b})$$

in the unknowns f_{21} and f_{12}, with solution

$$f_{21} = -\hat{m}/[1 + \exp(\hat{b})]$$

$$f_{12} = f_{21} \exp(\hat{b}).$$

Additional elementary compartmental models involving tracer kinetics can be found in [191], [218], [35], [29], [34], [194], [134].

9A. Tracer Concentration Equations

Due to the steady state conditions discussed earlier, we assume that V_i, the size of the ith compartment (see Section 25), and q_{iss}, the steady state amount of unlabeled material in the ith compartment, are both <u>constant</u> over time. This sets the stage for two different kinds of tracer studies.

In the first and probably most common type of experiment, the <u>concentration of tracer</u> in one or more compartments is followed and sampled over time. Thus it is preferable to have our basic tracer model in a form involving tracer concentrations rather than amounts. Since $V_i > 0$ is constant,

$$c_i(t) \equiv x_i(t)/V_i \qquad (i = 1, 2, \ldots, n)$$

gives the concentration of tracer in the ith compartment at time $t \geq 0$. Let V be the diagonal matrix diag $[V_1, V_2, \ldots, V_n]$, and let

$$c(t) \equiv [c_1(t), c_2(t), \ldots, c_n(t)]^T .$$

We introduce a change of variable into the linear compartmental model through the linear transformation x = Vc. Upon substitution of this product into the model, there results

$$V\dot{c} = \dot{x} = Ax + b = AVc + b,$$

so that the system of equations for the tracer concentrations is

(9.1)
$$\boxed{\dot{c} = (V^{-1}AV)c + V^{-1}b .}$$

Define the $n \times n$ matrix $M \equiv V^{-1}AV$; in component form the (i,j)-entry is thus

$$m_{ij} = a_{ij}V_j/V_i$$

for all $i, j = 1, 2, \ldots, n$. Note that M is a matrix which is similar to the compartmental matrix A, and they both share the same diagonal elements.

<u>Exercise 9.1.</u> Is M a compartmental matrix? State any properties that M has as a matrix.

9B. Tracer Specific Activity Equations

The second kind of experiment is where the <u>tracer specific activity</u> time course is followed. The specific activity of tracer in the ith compartment at time $t \geq 0$ is defined as

$$a_i(t) \equiv x_i(t)/q_{iss} \qquad (i = 1, 2, \ldots, n).$$

Analogous to our treatment in the previous case, introduce

$$Q \equiv diag [q_{1ss}, q_{2ss}, \ldots, q_{nss}]$$

as the $n \times n$ diagonal matrix of steady state tracee amounts, and the vector

$$a(t) \equiv [a_1(t), a_2(t), \ldots, a_n(t)]^T.$$

The appropriate change of variable for the linear compartmental equation in the present situation is $x = Qa$. This substitution leads to

$$Q\dot{a} = \dot{x} = Ax + b = AQa + b$$

or

$$(9.2) \qquad \boxed{\dot{a} = (Q^{-1}AQ)a + Q^{-1}b \, ,}$$

which will be known as the tracer specific activity equation. Equation (9.2) is convenient to use when the specific activity of the tracer is being observed [31].

It is now clear, from the fundamental linear compartmental model, or from variations (9.1) and (9.2), that we wish to <u>study linear continuous time dynamic systems in which the system matrix is either a compartmental matrix or a matrix which is (diagonally) similar to a compartmental matrix</u>. What we have here is an entire family of matrices $\{S^{-1}AS\}$ which are similar to a compartmental matrix A. Any choice of these matrices can be made to appear in the basic model by the simple change of variable $x = Sy$. Since such a fundamental change exists, all of these matrices should have some common property, and they do: They all share the same eigenvalues as A. We shall investigate these eigenvalues in Section 12.

SECTION 10.   ANALYTICAL SOLUTION OF THE TRACER MODEL

10A.   The General Solution of the Model

Our model for tracer kinetics in an n-compartment system is the set of first order linear differential equations

$$\dot{x}(t) = Ax(t) + b(t), \qquad t \geq 0,$$

(10.1)

$$x(0) = x_0,$$

where $x(t)$ is the $n \times 1$ state vector for the amounts of tracer in the compartments, $b(t)$ is the $n \times 1$ driving vector of tracer inputs, and A is an $n \times n$ compartmental matrix of constant fractional transfer coefficients.

From the theory of linear equations, it is well-known that the general solution of (10.1) is given by the general solution of the homogeneous portion

(10.2) $\dot{x} = Ax$

plus any particular solution of (10.1).   Let us for the moment consider (10.2) when $n = 2$.   Because (10.2) is a linear equation, a solution of the form

$$x(t) = [A_1 \exp(\lambda t) \quad A_2 \exp(\lambda t)]^T,$$

where A_i and λ are constants, is attempted.   Substitution of this x-vector into (10.2) leads to the linear algebraic system

$$(A-\lambda I)[A_1 \quad A_2]^T = 0.$$

This system has a <u>nontrivial</u> solution $[A_1 \quad A_2]^T$ if and only if $\det (A-\lambda I) = 0$; that is, provided λ is an <u>eigenvalue of A</u>.   For each of the two eigenvalues λ_j, $j = 1, 2$, of A, there is an associated nonzero solution $[A_{1j} \quad A_{2j}]^T$ which is an <u>eigenvector of A</u>.   Thus for each λ_j we have a solution

$$[A_{1j} \exp(\lambda_j t) \quad A_{2j} \exp(\lambda_j t)]^T.$$

Hence the general solution of (10.2) in the case $n = 2$ is

(10.3) $x(t) = \gamma_1 [A_{11} \quad A_{21}]^T \exp(\lambda_1 t) + \gamma_2 [A_{12} \quad A_{22}]^T \exp(\lambda_2 t)$

for scalars γ_1 and γ_2, which are determined by the initial condition x_0.

For the general case of n, the solution of (10.2) along with the initial condition $x(0) = x_0$, can be given in terms of the matrix exponential,

$$x(t) = \exp(tA)x_0.$$

This exponential can be constructed through diagonalization or, equivalently, by component matrices (see Section 20). Suppose the $n \times n$ compartmental matrix A is diagonalizable. Let S be the $n \times n$ matrix whose columns are the $n \times 1$ linearly independent right eigenvectors $e_1, \ldots, e_n$ corresponding to the eigenvalues $\lambda_1, \ldots, \lambda_n$ of A. The matrix exponential is then given by [153]

$$\exp(tA) = S \exp(t\Lambda) S^{-1}$$

where Λ is the $n \times n$ diagonal matrix $[\lambda_i \, \delta_{ij}]$, and

$$\exp(t\Lambda) = \text{diag} \{\exp(\lambda_1 t), \ldots, \exp(\lambda_n t)\}.$$

The matrix S is nonsingular since its columns are linearly independent. Moreover, the row vectors f_i^T of S^{-1} are the normalized left eigenvectors of A [153, p. 180]:

$$f_i^T A = \lambda_i f_i^T, \quad f_i^T e_i = 1, \quad i = 1, 2, \ldots, n.$$

(The component matrices of A are then $Z_i \equiv e_i f_i^T$, $i = 1, \ldots, n$.) Thus the solution $\exp(tA)x_0$, when the eigenvalues are distinct, can be written as

(10.4) $x(t) = \sum_{i=1}^{n} \gamma_i e_i \exp(\lambda_i t).$

The scalars γ_i are picked to match the initial condition $x(0) = x_0$. In fact, the constants are elements of the vector

$$\gamma \equiv S^{-1} x_0 \equiv [\gamma_1 \quad \gamma_2 \quad \cdots \quad \gamma_n]^T.$$

Form (10.4), and its particular case (10.3), is commonly called the <u>sum-of-exponentials model</u>.

Returning now to the nonhomogeneous equation (10.1), its complete solution is

(10.5)
$$x(t) = \exp(tA)x_0 + \int_0^t [\exp(t-\tau)A]b(\tau)d\tau \;,$$

where the integration is carried out element-wise [153, p. 119].

By the matrix $\exp(tA)$ in the above discussion, we also mean the sum of the series

(10.6) $I + tA + t^2A^2 / 2! + t^3A^3 / 3! + \ldots$

which converges for all values of t and for any matrix A [153, p. 116]. Usually it is preferable to compute $\exp(tA)$ by using eigenvalues and functions of a matrix [45] or by some other method [165] rather than attempt to sum the series (10.6).

Exercise 10.1.   Compute the matrix $\exp(tA)$ for the compartmental matrix

$$A = \begin{bmatrix} -1 & 1 \\ 1 & -1 \end{bmatrix}$$

and then write out in detail the model solution (10.5), using

$$x_0 \equiv [1 \quad 0]^T \text{ and } b \equiv [0 \quad 1]^T.$$

Exercise 10.2.   Verify directly that x(t) in (10.5) is a solution of the model (10.1).

It is worth noting that an important special case in (10.5) is when the tracer input vector b is _constant_ over time. Then the solution can be written as

(10.7) $x(t) = \exp(tA)x_0 - A^{-1}b$

assuming the matrix A is invertible.

10B. Nonnegativity of the Solution

It is clear from the solution form (10.5) that properties of the matrix exponential $\exp(tA)$ will determine the behavior of the solution vector x(t). In particular, we are able to guarantee the nonnegativity of x(t) for all $t \geq 0$, which is the only sign for x that physically makes sense.

Let us define a real $n \times n$ matrix $A \equiv [a_{ij}]$ to be essentially nonnegative if $a_{ij} \geq 0$ for all i, j, $i \neq j$ (compartmental matrices satisfy this condition). Such a matrix is also called a Metzler matrix [153], [76], [162]. These matrices arise in the study of economics, approximate solutions of parabolic partial differential equations [234, p. 257], and in the study of nuclear reactor theory [38]. The condition $a_{ij} \geq 0$, $i \neq j$, turns out to be equivalent to the fact that the system $\dot{x} = Ax$ preserves nonnegativity of the state vector x.

Exercise 10.3.   Prove that if A is a Metzler matrix, then exp(tA) is a nonnegative matrix for all $t \geq 0$. (A matrix $M \equiv [m_{ij}]$ is nonnegative provided $m_{ij} \geq 0$ for all i, j; we write $M \geq 0$.) Hint: Choose c such that $tA + cI \geq 0$ for some t, and consider the matrix exp(tA + cI).

Looking back to the general solution (10.5), we now observe that if $b(t) \geq 0$ for all $t \geq 0$ and if $x_0 \geq 0$, then $x(t) \geq 0$ for all $t \geq 0$ (also see [153, p. 204]). Hearon [123, p. 48] points out that $x(t) \geq 0$ for all $t \geq 0$ if and only if $a_{ij} \geq 0$ $(i \neq j)$ in the matrix A. Furthermore, for the solution x(t) to be bounded, it is necessary that $a_{ii} \leq 0$, as we have in our model.

Exercise 10.4.   Let $A \equiv [a_{ij}]$ be an $n \times n$ Metzler matrix such that $A^{-1} = A$.

(a) Give a nontrivial example of such an A. Must A be a compartmental matrix?

(b) If the system $\dot{x}(t) = Ax(t) + b$, $b > 0$, has a nonnegative equilibrium vector x_e, show that $-1 \leq a_{ii} \leq 0$ for all $i = 1, 2, \ldots, n$.

Exercise 10.5.   Differential equations, even when they are linear with constant coefficients, are often solved numerically [165]. Solve $\dot{x} = Ax$, $x(0) = x_0$, where

$$A = \begin{bmatrix} -0.5 & 0 \\ 0.5 & -0.25 \end{bmatrix} \qquad x_0 = \begin{bmatrix} 1 \\ 0 \end{bmatrix}$$

by a numerical integrator such as DVERK in the IMSL package [131]. In particular compute x(1), x(2), ..., x(10). Compare this with the analytical solution of the model. Confirm the nonnegativity of x(t).

11A.   The Connectivity Diagram

Many properties of our tracer model

$$\dot{x} = Ax + b, \qquad t \geq 0$$

(11.1)

$$x(0) = x_0$$

depend only on the disposition of zero and nonzero elements in the compartmental

matrix A (matrix singularity of A turns out to be such a property).   If $a_{ij} = 0$

($i \neq j$), then there is no flow of material from compartment j directly to compartment

i.   If $a_{ij} \neq 0$ ($i \neq j$), then material is passing directly from the jth to the ith

compartment.   A useful representation of structure of a compartmental system is the

connectivity diagram for the system which shows the nonzero transfers of material on

a directed graph.   The directed graph consists of a set of nodes (the compartments)

together with a set of directed edges (the flows) connecting certain of these nodes.

A compartmental system, complete with inputs and exits, and its associated connec-

tivity diagram are given in Figure 11.1.

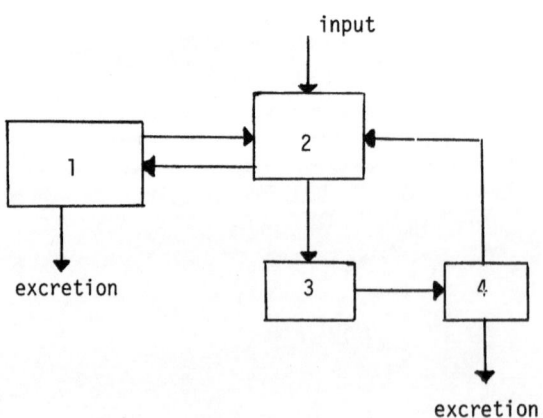

Figure 11.1. (a)

A Compartmental System

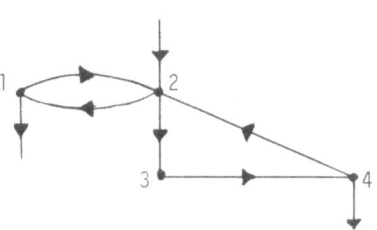

Figure 11.1. (b)

The Compartmental System Connectivity Diagram

11B. Common Compartmental Systems

There are some important types of connectivity in compartmental models that appear repeatedly in idealized physiological systems. The first is the catenary system in which the compartments connect in a series or chain as is shown in Figure 11.2. Only adjacent compartments communicate. Of special interest are catenary systems wherein there is tracer input into either compartment 1 or n, and single exit only from one of those two compartments. In particular, a common physiological catenary 3-compartment model is with the compartments identified as (1) blood plasma; (2) interstitial fluid; (3) cells. Here, due primarily to the kidney, the first compartment has the single exit to the outside. The compartmental matrix A for a catenary system as shown in Figure 11.2 is easily seen to be tridiagonal.

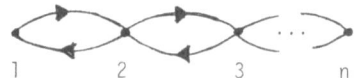

Figure 11.2

An n-Compartment Catenary System

Often useful in analyzing the kinetics of the distribution of a tracer injected into blood plasma and which enters the interstitial space is the mammillary system (see Figure 11.3). Here one compartment acts as the "mother" or central compartment and all other compartments are "daughters." Connectivity only takes place between the mother compartment and each individual daughter. It is standard to number the compartments commencing with the central compartment designated as number one. Pro-vided this numbering scheme is instituted, the compartmental matrix A has nonzero entries only in the first row, first column, and on the main diagonal.

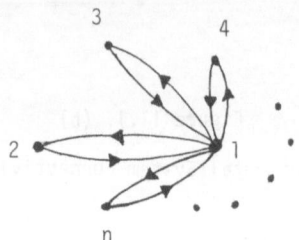

Figure 11.3.

An n-Compartment Mammillary System

An extension of the mammillary system concept is the <u>extended mammillary model</u> depicted in Figure 11.4 when n is odd.

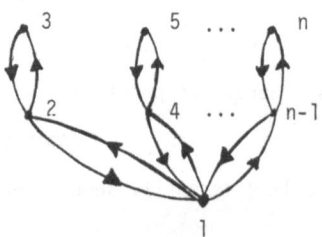

Figure 11.4.

The Extended Mammillary System

Finally, we mention a type of system very common in modeling - a <u>single exit compartmental (SEC) system</u> - in which in the model there is exactly one compartment from which excretion to the outside environment occurs [4], [8]. Call this exit compartment m. Then in the system matrix $A \equiv [a_{ij}]$, we have $a_{0m} > 0$, $a_{0j} = 0$ ($j \neq m$). SEC systems also carry the name <u>almost closed systems</u> [196]. Often SEC systems are combined with mammillary, catenary, or extended mammillary systems, in which there is tracer input into the central or first compartment, and excretion only from compartment 1, n, or the central compartment.

11C. Strongly Connected Systems

Now let us suppose that for a system S consisting of n compartments, the nodes (compartments) are numbered from 1 to n: $S \equiv \{1, 2, ..., n\}$. Suppose further that this ordering of compartments is such that there is a subset

$$T \equiv \{p, p+1, ..., n\}, \qquad 1 < p \leq n,$$

of S such that there is no flow of material from the compartments in T to the other compartments in S - T; that is, for each vertix j in T, the entry a_{ij} of A is zero whenever $i \notin T$. Such a set T of compartments is called a _trap_. The $n \times n$ system matrix A has the _partitioned matrix_ form

(11.2) $A = \left[\begin{array}{c|c} B & 0 \\ \hline D & C \end{array}\right]$

where B is a $(p-1) \times (p-1)$ matrix, C is a square matrix, and 0 is a matrix of zeros. An example is given in Figure 11.5. In this case, p = 4 and T = {4,5} is a trap.

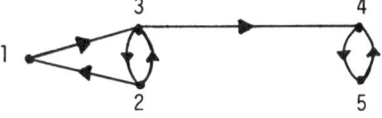

$$\left[\begin{array}{ccc|cc} a_{11} & a_{12} & 0 & 0 & 0 \\ 0 & a_{22} & a_{23} & 0 & 0 \\ a_{31} & a_{32} & a_{33} & 0 & 0 \\ \hline 0 & 0 & a_{43} & a_{44} & a_{45} \\ 0 & 0 & 0 & a_{54} & a_{55} \end{array}\right]$$

Figure 11.5

A Compartmental System Which Has a Trap,
and Its System Matrix

Compartmental matrices which can be brought into the partitioned form (11.2) through a renumbering of nodes or compartments are <u>reducible matrices</u>; otherwise the matrix is <u>irreducible</u>. If, in starting from <u>any</u> compartment, material can pass to <u>any</u> other compartment in the system by some path, then the system and its associated connectivity diagram are said to be <u>strongly connected</u>. It is clear from the preceding discussion that strong connectivity of a compartmental system is equivalent to saying that the system matrix A is irreducible (also see [143, p. 281]).

One of the values of recognizing that a given square matrix A is reducible is in the advantage it gives when solving the linear algebraic equation $Ax = b$, where b is a prescribed vector of appropriate dimension. For if we partition A as in (11.2) and use corresponding partitions in the vectors x and b, then the equation $Ax = b$ splits into

$$Bx_1 = b_1, \qquad Dx_1 + Cx_2 = b_2.$$

Thus to solve $Ax = b$, one first solves the lower dimensional equation $Bx_1 = b_1$ and then uses that solution x_1 in the second equation to determine x_2. A similar advantage is to be had in dealing with the solution of the algebraic eigenvalue problem $Av = \lambda v$.

For further comments and study on connectivity and related graph theory topics, see [187], [175], [114], [206].

SECTION 12.   SYSTEM EIGENVALUES AND STABILITY

12A. Nonpositive Eigenvalues

The subject of this section is that of eigenvalues and eigenvectors of a compartmental matrix A. As we have seen, the solution of the model

(12.1)
$$\dot{x}(t) = A\,x(t) + b(t), \qquad t \geq 0,$$
$$x(0) = x_0,$$

can be decomposed into a sum of terms involving eigenvectors associated with various eigenvalues.

Let us start with a discussion of the eigenvalues of an $n \times n$ compartmental matrix A. Suppose these eigenvalues are ordered in terms of magnitudes as

(12.2) $|\lambda_1| \geq |\lambda_2| \geq \cdots \geq |\lambda_n|$.

In virtually all applications of compartmental modeling, these eigenvalues are real and distinct numbers. In any case, we can always assert the following [121, p. 139]. (As a consequence, one can show that det $(-A) \geq 0$ [123, p. 49].)

THEOREM 12.1.   The real part of any eigenvalue of A is nonpositive. Moreover, the matrix has no purely imaginary eigenvalues.

Proof.   The Gerschgorin Circle Theorem for eigenvalues of any square matrix A states that if the circular disk D_k in the complex plane is defined by

$$D_k \equiv \{z: \; |z - a_{kk}| \leq \sum_{i \neq k} |a_{ik}|\}, \qquad k = 1,\, 2,\, \ldots,\, n,$$

then all of the eigenvalues of A are contained in the union of the disks $\{D_k\}$ [222], [143]. (The notation $i \neq k$ on the above summation means that we sum over all $i = 1,\, 2,\, \ldots,\, n$ except for $i = k$.) In particular, if A is a compartmental matrix, we have $a_{ik} \geq 0$ for all $i \neq k$, and since the sum of the general k^{th} column is $-a_{0k}$, then

$$\sum_{i \neq k} a_{ik} = -a_{0k} - a_{kk} \leq -a_{kk} = |a_{kk}|$$

for each k = 1, 2, ..., n. Hence for each k, the disk D_k is contained in the circular region $|z - a_{kk}| \leq |a_{kk}|$. Since any eigenvalue is in at least one of these regions, then the real part of that eigenvalue is in the left half plane. Furthermore, because these regions are disks, no eigenvalue can lie on the imaginary axis (except zero, if it is an eigenvalue).

///

Exercise 12.1. Show that if every compartment in the system has an exit to the outside environment, then zero is not an eigenvalue of A, and hence A is nonsingular. On the other hand, prove that if no compartment in the system has an exit, then there is at least one zero eigenvalue of A and as a consequence A is a singular matrix.

Note that since no purely imaginary eigenvalues of A exist, oscillations (if any) which occur in the solution of the model (12.1) are necessarily damped.

12B. The Smallest Magnitude Eigenvalue

Compartmental matrices are closely related to nonnegative matrices, that is, matrices M all of whose elements are nonnegative. We write $M \geq 0$. For instance, take any n × n compartmental matrix A. Then for any constant c such that

$$c \geq \max_{1 \leq i \leq n} |a_{ii}| \, ,$$

then $A + cI \geq 0$. One important property of a nonnegative matrix concerns its dominant eigenvalue. For any square matrix M, let

$$\rho(M) \equiv \max \{|r| : r \text{ is an eigenvalue of } M\} \, .$$

This number $\rho(M)$ is called the spectral radius of M. If $M \geq 0$, the Perron-Frobenius Theorem says that M has a real nonnegative eigenvalue which is equal to $\rho(M)$, and a corresponding nonnegative eigenvector v [143]. Thus for our matrix $A + cI \geq 0$, we have that $r \equiv \rho(A + cI) \geq 0$ is an eigenvalue of A + cI with a corresponding eigenvector $v \geq 0$.

Exercise 12.2.   Show that the eigenvalues of A + cI are λ_i + c, where λ_i are the eigenvalues of A.

Due to this exercise, we have λ_ℓ + c = r for some eigenvalue λ_ℓ of the compartmental matrix A. Thus

(12.3) $\lambda_\ell + c = \rho(A + cI) \geq |\lambda_i + c|$

for all i = 1, 2, ..., n. Furthermore,

(12.4) $|\lambda_i - (-c)| \geq \| |\lambda_i| - |c| \| = \| |\lambda_i| - c |$.

From (12.3) combined with (12.4), it follows that

$$\lambda_\ell + c \geq |\lambda_i| - c \geq - \lambda_\ell - c$$

for all i = 1, 2, ..., n, since λ_ℓ is now known to be real and nonpositive. The last inequality says that

$$|\lambda_i| \geq |\lambda_\ell| \qquad\qquad i = 1, 2, ..., n.$$

Hence λ_ℓ is an eigenvalue of A of smallest magnitude; since one such eigenvalue is designated as λ_n in (12.2), we have proved part of the next theorem (see [153, p. 205] or [123, p. 44]).

THEOREM 12.2.   For any n × n compartmental matrix A, with eigenvalues as in (12.2), the eigenvalue λ_n is real and nonpositive. To that eigenvalue there corresponds an eigenvector v $\geq$ 0. If λ_i is any other eigenvalue of A, then Re(λ_i) $\leq \lambda_n$.

Many compartmental systems are strongly connected so that the compartmental matrix is irreducible. Then additional results are available about λ_n. The proof of the following theorem for other than compartmental matrices is found in [234, p. 258] and is based on the Perron-Frobenius theory (also see [143]).

THEOREM 12.3.   Suppose a system of n compartments is strongly connected. Then the real eigenvalue λ_n of the associated compartmental matrix A is such that

(a)   to λ_n there corresponds a <u>positive</u> eigenvector v;

(b) λ_n is a <u>simple</u> eigenvalue of A;

(c) if λ_i is any other eigenvalue of A, then $\text{Re}(\lambda_i) < \lambda_n$;

(d) λ_n increases when any element of A increases.

Theorem 12.3 tells us that the slowest transient in a strongly connected compartmental system corresponds to the simple real root λ_n.

12C. Symmetrizable Compartmental Matrices and Real Eigenvalues

We have already said that in most applications of compartmental analysis, the eigenvalues are real (and nonpositive). The previous theorems tell us that λ_n is always real. When are other eigenvalues real? One case that leads to real eigenvalues is the case where "symmetry" is involved.

For any two complex n-dimensional vectors $x \equiv [x_i]$, $y \equiv [y_i]$, the usual <u>inner product</u> is defined by

$$\langle x, y \rangle \equiv \sum_{i=1}^{n} x_i \bar{y}_i$$

where the bar denotes conjugation of the complex number y_i. For any $n \times n$ <u>Hermitian matrix</u> $G \equiv [g_{ij}]$, that is, where $\bar{g}_{ij} = g_{ji}$ for all i, j, we have the <u>self-adjoint</u> condition

$$(12.5) \qquad \langle x, Gy \rangle = \sum_{i=1}^{n} x_i \sum_{j=1}^{n} \bar{g}_{ij} \bar{y}_j = \sum_{i=1}^{n} g_{ji} x_i \sum_{j=1}^{n} \bar{y}_j = \langle Gx, y \rangle .$$

Let us now suppose that for a given $n \times n$ real matrix A there exists a real symmetric $n \times n$ matrix $Q \equiv [q_{ij}]$ such that

(a) QA is a symmetric matrix;

(b) Q is a <u>positive definite matrix</u>, i.e., Q possesses the property that $\langle x, Qx \rangle > 0$ for any nonzero vector x.

Let λ be an eigenvalue of A with associated (nonzero) eigenvector v. Then because Q is symmetric, equation (12.5) tells us that

$$(12.6) \qquad \langle Q(Av), v \rangle = \langle Av, Qv \rangle .$$

Moreover, $Av = \lambda v$ yields

$$(12.7) \qquad <Av, Qv> = <\lambda v, Qv> = \lambda <v, Qv> \ .$$

In equation (12.7), the factor $<v, Qv>$ is positive since Q is a positive definite matrix. Also

$$(12.8) \qquad <(QA)v, v> = <v, (QA)v> = \overline{<(QA)v, v>}$$

by condition (a) and result (12.5). Thus $<QAv, v>$ is a real number. Equality of (12.6) and (12.7) now result in the fact that λ must be real.

THEOREM 12.4.   If there exists, for a given $n \times n$ matrix A, an $n \times n$ symmetric positive definite square matrix Q such that QA is also symmetric, then the eigenvalues of A are real. Moreover, eigenvectors corresponding to distinct eigenvalues of A are orthogonal with respect to the Q-inner product $<x, y>_Q \equiv <x, Qy>$.

An $n \times n$ matrix A that fulfills the hypothesis of this theorem is called a symmetrizable matrix. The conditions $QA = (QA)^T$ and $Q = Q^T$ imply the general constraint that the elements q_{ij} of Q must satisfy the equations

$$(12.9) \qquad \sum_{k=1}^{n} q_{ik} a_{kj} = \sum_{k=1}^{n} a_{ki} q_{kj}$$

for all $i, j = 1, 2, \ldots, n$.

Now we ask an important question. Since we are interested in real eigenvalues for our compartmental systems, just what types of systems have these real eigenvalues?

Let $A \equiv [a_{ij}]$ be a catenary compartmental matrix with the usual numbering of compartments so that $a_{ij} = 0$ for all i, j such that $|i - j| > 1$. In this case we can construct an appropriate $n \times n$ matrix Q which is diagonal:

$$Q \equiv \text{diag} \{q_{11}, q_{22}, \ldots, q_{nn}\} \ ,$$

where the positive definite condition on Q requires that $q_{ii} > 0$ for all i. Then (12.9) reduces to

(12.10) $\qquad q_{ii}a_{ij} = a_{ji}q_{jj}$ $\qquad\qquad i, j = 1, 2, \ldots, n.$

We need only consider the case when $j = i - 1$ and catenary systems for which

(12.11) $\qquad a_{i,i-1} \neq 0,$ $\qquad\qquad i = 2, 3, \ldots, n.$

Then from (12.10), q_{ii} is given recursively by

$$q_{ii} = a_{i-1,i}\, q_{i-1,i-1} / a_{i,i-1}$$

for all $i = 2, 3, \ldots, n$, if we provide a definition for q_{11}. It is simplest to choose $q_{11} \equiv 1$. Then

$$q_{22} = a_{12}/a_{21}$$

$$q_{33} = a_{12}a_{23}/a_{32}a_{21}$$

$$q_{44} = a_{12}a_{23}a_{34}/a_{43}a_{32}a_{21}$$

$$\vdots \qquad\qquad \vdots$$

THEOREM 12.5. A catenary n-compartmental system with compartments ordered such that the compartmental matrix $A \equiv [a_{ij}]$ satisfies (12.11) is symmetrizable.

Exercise 12.3. Show that a mammillary compartmental matrix is symmetrizable.

Let $A \equiv [a_{ij}]$ be an $n \times n$ compartmental matrix with connectivity diagram Z. By a circuit in Z, we mean an ordered set of k $(\leq n)$ distinct indices $\{j_1, j_2, \ldots, j_k\}$ such that Z contains the directed edges

$$\overrightarrow{j_1 j_2}, \ \overrightarrow{j_2 j_3}, \ \ldots, \ \overrightarrow{j_k j_1} \ .$$

Note that this is equivalent to saying that the product of matrix elements

$$a_{j_1 j_k} \cdots a_{j_4 j_3}\, a_{j_3 j_2}\, a_{j_2 j_1} \neq 0 \ .$$

The length of the circuit is said to be k.

THEOREM 12.6. [140]. If the longest circuit in the diagram Z for a square matrix A is of length two, then each eigenvalue of A is real.

As an example, consider the compartmental system whose connectivity diagram is in Figure 12.1. The circuit from compartment 2 to 3 and back to 2 is of length two; the same is true for the circuit from 5 to 6. Since these are the longest circuits in the system, we can conclude via Theorem 12.6 that all eigenvalues of the asso- ciated compartmental matrix

$$
A = \begin{bmatrix}
a_{11} & 0 & 0 & 0 & 0 & 0 \\
a_{21} & a_{22} & a_{23} & 0 & 0 & 0 \\
0 & a_{32} & a_{33} & 0 & 0 & 0 \\
0 & 0 & 0 & a_{44} & a_{45} & 0 \\
0 & a_{52} & 0 & 0 & a_{55} & a_{56} \\
0 & 0 & 0 & 0 & a_{65} & a_{66}
\end{bmatrix}
$$

are real.

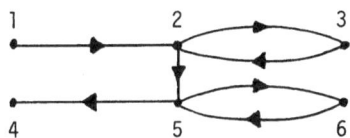

Figure 12.1.

A Connectivity Diagram in Which No Circuits
Are of Length Greater Than Two

Cases of special interest, which fall under the conditions of Theorem 12.6, are the catenary system and the mammillary system. Hearon [123, p. 54] discusses other classes of systems whose compartmental matrix A is symmetrizable and therefore has only real eigenvalues. For additional information on symmetrizability, see S. V. Parter and J. W. T. Youngs, The symmetrization of matrices by diagonal matrices, J. Math. Analy. Appl. 4 (1962), 102-110.

12D. Distinct Eigenvalues

In compartmental modeling there is also interest in the case when the eigen-values are <u>distinct</u>. If $A \equiv [a_{ij}]$ is a mammillary compartmental matrix, then it follows from the last theorem that the eigenvalues must be real. Moreover, Sheppard and Householder [211, p. 514] have shown that the eigenvalues of a mammillary matrix are distinct provided

$$a_{1i} \neq a_{1j} \qquad \text{for all } i, j, \qquad i \neq j \neq 1,$$

$$a_{1i}a_{i1} > 0 \qquad \text{for } i = 1, 2, \ldots, n$$

(the mother compartment is designated the first compartment). More can be said about the <u>separation</u> of the mammillary case eigenvalues. For convenience, suppose now that the mother compartment is numbered n and the daughter compartments are such that a_{ni}, $a_{in} > 0$ for all $i < n$. The form of this mammillary matrix is then

$$A = \begin{bmatrix} a_{11} & 0 & \cdots & a_{1n} \\ 0 & a_{22} & \cdots & a_{2n} \\ \cdot & \cdot \cdot \cdot & \cdot \cdot \cdot & \cdot \\ a_{n1} & a_{n2} & \cdots & a_{nn} \end{bmatrix}.$$

Remove the last row and last column of A. The resulting $(n-1) \times (n-1)$ matrix

$$\text{diag } \{a_{11}, a_{22}, \ldots, a_{n-1,n-1}\}$$

has the $\{a_{ii}\}$ as eigenvalues. Furthermore, if we let

$$\gamma_1 \equiv \min_{1 \leq i \leq n-1} \{a_{ii}\} \equiv a_{i_1 i_1}$$

$$\gamma_2 \equiv \min_{\substack{1 \leq i \leq n-1 \\ i \neq i_1}} \{a_{ii}\} \equiv a_{i_2 i_2}$$

$$\begin{matrix} \cdot & \cdot & \cdot \\ \cdot & \cdot & \cdot \\ \cdot & \cdot & \cdot \end{matrix}$$

then the γ_i s separate the eigenvalues λ_i of A in the sense that [222, p. 259]

$$\lambda_1 \leq \gamma_1 \leq \lambda_2 \leq \gamma_2 \leq \lambda_3 \leq \cdots \leq \lambda_{n-1} \leq \gamma_{n-1} \leq \lambda_n .$$

Exercise 12.4.   Verify this last result in the case of the SEC mammillary matrix

$$A = \begin{bmatrix} -7 & 0 & 0 & 3 \\ 0 & -4 & 0 & 2 \\ 0 & 0 & -2 & 2 \\ 7 & 4 & 2 & -8 \end{bmatrix}$$

using a subroutine such as EIGRF in the IMSL package [131].

It can be shown that a <u>catenary</u> compartmental matrix $A \equiv [a_{ij}]$ has distinct eigenvalues if

$$a_{i,i+1} \neq 0, \quad a_{i+1,i} \neq 0, \quad 1 \leq i \leq n$$

(see [96, p. 30]).

The importance of the realness and distinctness of the eigenvalues λ_i of a compartmental matrix A is that the solution of the model $\dot{x} = Ax + b$ then assumes a form that is constantly being met in actual tracer studies - namely as the sum-of-exponentials. As we can see from the analytic solution of the homogeneous part of the model, the solution for the ith component (i = 1, 2, ..., n) is

$$(12.12) \quad x_i(t) = \sum_{j=1}^{n} A_{ij} \exp(\lambda_j t)$$

where the A_{ij} s are constants. These full solution forms (12.12) can be curve-fitted to real-valued data since the eigenvalues λ_i are real and distinct.

12E. Compartmental Model Stability

Now consider the model $\dot{x}(t) = Ax(t) + b$ for tracer kinetics in an open system for which the tracer input vector b is <u>constant</u> and A is an <u>invertible</u> compartmental matrix (if the system was closed, then A would be a singular matrix; we will discuss conditions under which A is invertible in the next section). This constant tracer input b will drive the system to steady state with respect to the tracer. For, if there is a complete complement of eigenvectors $\{v_i\}$ of A, then the general solution of the model is of the form (for constants c_k)

$$(12.13) \qquad x(t) = \sum_{k=1}^{n} c_k \exp(\lambda_k t) \, v_k - A^{-1}b$$

where $Re(\lambda_k) < 0$ for all k. Writing each eigenvalue in terms of its real and imaginary parts, $\lambda_k = \gamma_k + i\,\omega_k$, it is immediately seen that

$$|c_k \exp(\lambda_k t)| = |c_k| \exp(\gamma_k t).$$

Since the real part γ_k determines the rate of growth of the exponential, and since (12.13) holds with $\gamma_k < 0$, it is then clear that

$$x(t) \to x_e \equiv - A^{-1}b, \qquad t \to \infty,$$

for any initial tracer dose x(0). This convergence of x(t) to x_e as $t \to \infty$ says that the model is <u>asymptotically stable</u> [222], [153].

The tracer equilibrium x_e is such that $0 = \dot{x}_e = Ax_e + b$. One use of the constant equilibrium vector x_e is the following. If x is a solution of $\dot{x} = Ax + b$, and z is defined by $z \equiv x - x_e$, then z satisfies the <u>homogeneous</u> system $\dot{z} = Az$, $z(0) = x(0) - x_e$.

Because the eigenvalue λ_n of A is dominant in the sense that

$$\gamma_k \le \gamma_n < 0 \qquad k = 1, 2, \ldots, n-1,$$

then the term of (12.13) containing λ_n will affect the trajectory of the solution curve for the longest length of time. In fact, for moderate to large values of t, the solution (12.13) is approximated by

$$(12.14) \qquad c_n \exp(\lambda_n t) \, v_n - A^{-1}b.$$

12F. Bounds on the Extreme Eigenvalues

Because of the importance of λ_n on the solution of the model, we are encouraged to find some easily computable estimates on the magnitude of this eigenvalue.

THEOREM 12.7. [204], [123].   Let $A \equiv [a_{ij}]$ be any $n \times n$ compartmental matrix for which S and s are the largest and smallest row sums, respectively, and as usual

$$a_{0j} = - \sum_{i=1}^{n} a_{ij} \qquad (j = 1, 2, \ldots, n).$$

Then the eigenvalue λ_n of A of smallest magnitude satisfies

(a) $s \leq \lambda_n \leq \min\{0, S\}$;

(b) $|\lambda_n| \leq \min_{1 \leq i \leq n} |a_{ii}|$;

(c) $\min_{1 \leq j \leq n} a_{0j} \leq |\lambda_n| \leq \max_{1 \leq j \leq n} a_{0j}$.

Proof.   As has been previously shown, the matrix $A + cI \geq 0$ for $c \equiv \max|a_{ii}|$.   Now by the "standard continuity argument," we mean the following.   Replace every zero entry in the matrix $A + cI$ by some prescribed $\varepsilon > 0$.   Then the new matrix so constructed will have all positive elements.   All results used below will then hold as $\varepsilon \to 0^+$.

By the standard continuity argument we may assume that $A + cI > 0$.   Then from Perron-Frobenius theory, we can conclude that there is an eigenvector $w \equiv [w_1 \ w_2 \ \ldots \ w_n]^T > 0$ corresponding to the dominant eigenvalue $\lambda_n + c = \rho(A + cI)$.   Let

$$w_\ell \equiv \max_i w_i, \quad w_t \equiv \min_i w_i > 0.$$

Now equation $(A + cI)w = (\lambda_n + c)w$ implies

(12.15) $$\sum_{j \neq i} a_{ij} w_j = (\lambda_n - a_{ii}) w_i$$

for all $i = 1, 2, \ldots, n$.   If $i = \ell$, then we observe from (12.15) that

$$\left(\sum_{j \neq \ell} a_{\ell j} \right) w_\ell \geq (\lambda_n - a_{\ell\ell}) w_\ell$$

or that

$$S \geq \sum_{j=1}^{n} a_{\ell j} \geq \lambda_n .$$

Exercise 12.5. In a similar fashion, show that

$$\lambda_n \geq \sum_{j=1}^{n} a_{tj} \geq s.$$

To continue with the proof of Theorem 12.8, we observe that since $w_j > 0$ for all j, then $0 \leq \lambda_n - a_{ii}$ for i = 1, 2, ..., n, from equation (12.15). Thus result (b) obtains.

Exercise 12.6. Apply the above stated arguments to the transpose of the matrix A + cI to prove part (c) of the Theorem.

Exercise 12.7. To complement part (b) in the last Theorem, it is conjectured that the largest magnitude eigenvalue λ_1 of a compartmental matrix A satisfies

$$(12.16) \qquad |\lambda_1| \geq \max_{1 \leq i \leq n} |a_{ii}|.$$

Prove or disprove this conjecture. Let m be such that $|a_{mm}| = \max_i |a_{ii}| = c$. Then show that

$$(12.17) \qquad 2c - a_{0m} \geq |\lambda_1| \geq \sum_{i=1}^{n} |a_{ii}|/n.$$

There is a case for which (12.16) has been verified. The real square matrix $A \equiv [a_{ij}]$ is sign-symmetric if $a_{ij}a_{ji} \geq 0$ and zero off-diagonal elements are symmetrically disposed (i.e., $a_{ij} = 0$ if and only if $a_{ji} = 0$). Suppose now that the compartmental matrix A has no circuits of length greater than two, and is sign-symmetric. Then A is symmetrizable [123, p. 53] and there is a real symmetric matrix $B \equiv [b_{ij}]$ which is similar to A such that

$$b_{ij} = (a_{ij}a_{ji})^{\frac{1}{2}}, \quad i \neq j; \quad b_{ii} = a_{ii}.$$

It is easily shown that for a real symmetric matrix B with eigenvalues $\lambda_1 \leq \lambda_2 \leq \cdots \leq \lambda_n$, the inequalities $\lambda_1 \leq b_{ii} \leq \lambda_n$ hold. Since the diagonal elements of B are identical to those of A, and because the λ_i are also the eigenvalues of A, we then have (12.16).

To demonstrate some of these bounds, consider the compartmental matrix

$$(12.18) \quad A = \begin{bmatrix} -0.72 & 0.0 & 0.1 & 0.3 & 0.2 \\ 0.1 & -2.0 & 0.0 & 0.0 & 0.2 \\ 0.0 & 1.0 & -0.3 & 0.0 & 0.2 \\ 0.4 & 0.6 & 0.2 & -0.4 & 0.7 \\ 0.2 & 0.4 & 0.0 & 0.1 & -1.3 \end{bmatrix}$$

for which $S = 1.5$, $s = -1.7$, $a_{01} = 0.02$, $\min |a_{ii}| = 0.3$, and $\max |a_{ii}| = 2.0$. Hence from Theorem 12.7,

(a) $-1.7 \leq \lambda_5 \leq 0$;

(b) $|\lambda_5| \leq 0.3$;

(c) $0 \leq |\lambda_5| \leq 0.02$;

and $4 \geq |\lambda_1| \geq 0.944$ from (12.17). In fact, the eigenvalues of A are

$$\lambda_1 = -2.091 \qquad\qquad \lambda_4 = -3.597 \times 10^{-1}$$

$$\lambda_2 = -1.210 \qquad\qquad \lambda_5 = -5.100 \times 10^{-3},$$

$$\lambda_3 = -1.052$$

as computed using the IMSL package [131].

We must be careful when using computer subroutines to calculate eigenvalues of matrices. This is especially true for SEC matrices. For this class of compartmental matrices, property (c) of Theorem 12.7 reduces to

$$(12.19) \quad 0 \leq |\lambda_n| \leq a_{0m},$$

where the system has exit only from the mth compartment. The potential trouble arises from the fact that the excretion rate a_{0m} may be quite small. As a concrete example, consider the SEC matrix

$$A = \begin{bmatrix} -2-a_{01} & 4 & 0 & 1 \\ 0 & -10 & 0 & 0 \\ 1 & 5 & -1 & 1 \\ 1 & 1 & 1 & -2 \end{bmatrix}$$

in which $a_{01} = 10^{-5}$. Thus $-10^{-5} \le \lambda_4 \le 0$ from (12.19). Hence the matrix is "close" to being singular. Using subroutine EIGRF in the IMSL package, the eigenvalues were reported as -10, -3, -2, and 0. Because zero, to within the accuracy requested, is an eigenvalue, we might conclude that A is singular. This is a false conclusion, however, since $\det A = 10^{-4} \ne 0$.

Exercise 12.8. Let the characteristic equation of an $n \times n$ compartmental matrix A be

$$s^n + s_1 s^{n-1} + \ldots + s_n = 0$$

(see (16.17)). Show that any eigenvalue λ_i of A satisfies

$$-\lambda \le Re(\lambda_i) \le 0$$

where λ is the unique positive root of the polynomial

$$p(x) \equiv x^n - Re(s_1)x^{n-1} - |s_2|x^{n-2} - \ldots - |s_n| = 0.$$

See E. Deutsch, Bound on zeros of a polynomial, Amer. Math. Monthly 87 (1980), 496, and Anderson [3].

SECTION 13.   THE INVERSE OF A COMPARTMENTAL MATRIX

13A. Invertibility Conditions

In the last section we saw that under the condition of a __constant input__ vector b, the compartmental model

(13.1) $\dot{x}(t) = Ax(t) + b$

comes to a natural resting spot or equilibrium provided A^{-1} exists. A constant vector x_e is an __equilibrium point__ of model (13.1) provided it has the property that once the state vector is equal to x_e it remains equal to that vector for all future time [153]. For the system (13.1), an equilibrium point x_e satisfies the equation

(13.2) $0 = Ax_e + b$.

Hence if A is nonsingular, there is a __unique__ equilibrium point $x_e = - A^{-1}b$ for the system; if A is singular there may or may not be equilibrium points. Thus it is important to know under what conditions a compartmental matrix A is invertible.

Invertibility of A can of course be given in terms of its eigenvalues. In particular, if the smallest magnitude eigenvalue $\lambda_n \neq 0$, then A^{-1} exists. Moreover, a necessary and sufficient condition that all eigenvalues of - A have positive real parts is that the leading principal minors of - A be positive [123, p. 50]. But more can be said using the Perron-Frobenius theory since we know that all eigenvalues of the matrix A lie in the left half of the complex plane.

THEOREM 13.1.   [153, p. 206]. Let A be a Metzler matrix. Then A^{-1} exists and consists of all nonpositive entries (we write $A^{-1} \leq 0$) if and only if all of the eigenvalues of A lie strictly within the left half of the complex plane.

Invertibility of a compartmental matrix can also be given in terms of the connectivity of the compartmental system. It is known that strong connectivity of the compartmental system (or equivalently, irreducibility of the compartmental matrix) is __sufficient__ to say that the associated compartmental matrix is nonsingular [123], [226], [234]. However, this condition is not necessary. For example, the SEC matrix,

$$A = \begin{bmatrix} -2 & 1 & 0 & 0 \\ 1 & -1 & 1 & 1 \\ 0 & 0 & -1 & 0 \\ 0 & 0 & 0 & -1 \end{bmatrix}$$

with connectivity diagram in Figure 13.1, is invertible,

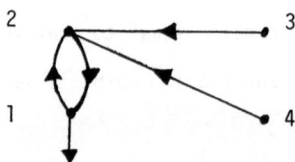

Figure 13.1.

but the system is not strongly connected.

In what is forthcoming, we need consider only <u>open compartmental systems</u>. A compartmental matrix associated with a closed compartmental system (i.e., all column sums of A are zero) is singular due to the following exercise.

<u>Exercise 13.1</u>.   Let M be any square matrix of constants with column sums (or row sums) <u>all</u> equal to the value λ. Then λ is an eigenvalue of M. Moreover, if $M \geq 0$, then λ is an eigenvalue of maximal modulus.

Suppose now that an open n-compartment system is denoted by $S \equiv \{1, 2, \ldots, n\}$ and the system has compartmental matrix $A \equiv [a_{ij}]$. Let

(13.3) $T \equiv \{p, p+1, \ldots, n\}$

for some $p \leq n$. Recall that T is a <u>trap</u> if and only if $a_{ij} = 0$ for all i, j such that $j \in T$ and $i \notin T$. Thus a trap is a subsystem of S with no excretion to things outside itself. Let us assume the tracer input is a bolus injection into one or more of the compartments at time zero. Then the tracer kinetics can be given by equation (13.1) with b = 0, but with $x(0) \neq 0$. The system having no traps is equivalent to stating that the amount of tracer $x_i(t)$ in the ith compartment, $i = 1$, 2, ..., n, goes to zero as $t \to \infty$. But $x_i \to 0$ as $t \to \infty$ for all i is necessary and sufficient to tell us that the real part of any eigenvalue of A is negative [143, Chapter 8]. Hence we have the following theorem which is discussed in detail in [97] for general compartmental systems, and in [8], [4] for the SEC case.

THEOREM 13.2.   Let S be an open n-compartment system with associated compartmental matrix A.   Then A is invertible if and only if S contains no traps.

The previous theorem can be shown to hold true without appealing directly to the theory of differential equations. This is done by treating three cases. (Suppose the compartments are so numbered that any trap present is given as in (13.3).)

Exercise 13.2.   In the first case, suppose p = n so that T = {n}. Prove that A is a singular compartmental matrix.

The second case is the opposite extreme. Assume the trap T contains all compartments except the first, i.e., T = {2, 3, ..., n}. Let us illustrate the proof when n = 4. The compartmental matrix must take the form

$$(13.4) \quad A = \begin{bmatrix} a_{11} & 0 & 0 & 0 \\ a_{21} & a_{22} & a_{23} & a_{24} \\ a_{31} & a_{32} & a_{33} & a_{34} \\ a_{41} & a_{42} & a_{43} & a_{44} \end{bmatrix} \qquad \begin{array}{l} a_{22} = -a_{32} - a_{42} \\ \\ a_{33} = -a_{23} - a_{43} \\ \\ a_{44} = -a_{24} - a_{34} \end{array}$$

in which each $a_{0j} = 0$ since $j \in T$. Now

$$\det A = a_{11} \cdot \det \begin{bmatrix} a_{22} & a_{23} & a_{24} \\ a_{32} & a_{33} & a_{34} \\ a_{42} & a_{43} & a_{44} \end{bmatrix}$$

However, the second determinant is zero since each of its column sums is zero due to the conditions of (13.4) and utilizing Exercise 13.1. Hence A is singular.

The final case is when $2 < p < n$ in (13.3). Again, for purposes of illustration, we take $n = 4$. Let us suppose the trap is $T = \{3, 4\}$. Hence the system matrix is

(13.5)
$$A = \begin{bmatrix} a_{11} & a_{12} & 0 & 0 \\ a_{21} & a_{22} & 0 & 0 \\ a_{31} & a_{32} & a_{33} & a_{34} \\ a_{41} & a_{42} & a_{43} & a_{44} \end{bmatrix}$$

in which $a_{33} = -a_{43}$ and $a_{44} = -a_{34}$ since $a_{03} = a_{04} = 0$ due to the presence of the trap. At this point in the discussion, we review a method of evaluating an $n \times n$ determinant in n steps which suits our needs very nicely and can be quite valuable as a computing device.

THEOREM 13.3. [193]. If $[a_{ij}]$ is any $n \times n$ matrix and $a_{11} \neq 0$, then $\det [a_{ij}] = \det [b_{ij}]/a_{11}^{n-2}$, where $[b_{ij}]$ is the $(n-1) \times (n-1)$ matrix whose (i, j)-entry is the 2×2 determinant

$$\begin{vmatrix} a_{11} & a_{1(j+1)} \\ a_{(i+1)1} & a_{(i+1)(j+1)} \end{vmatrix}.$$

For the compartmental matrix in (13.5), we first note that $a_{11} \neq 0$ since $1 \notin T$. Thus, according to Theorem 13.3,

$$\det A = \frac{1}{a_{11}^2} \begin{vmatrix} x & 0 & 0 \\ x & -a_{11}a_{43} & a_{11}a_{34} \\ x & a_{11}a_{43} & -a_{11}a_{34} \end{vmatrix}$$

where the x s denote entries in which we have no interest. If we factor the common expression $a_{11}a_{43}$ out of the second column of the matrix, and also factor $-a_{11}a_{34}$ out of the third column, then the last expression simplifies to

$$\det A = -a_{43}a_{34} \cdot \det \begin{bmatrix} x & 0 & 0 \\ x & -1 & -1 \\ x & 1 & 1 \end{bmatrix}.$$

Since two columns of the last determinant are identical, $\det A = 0$, and the proof is complete.

13B. A Neumann Series for the Inverse Matrix

Theorem 13.1 contains not only information about when an $n \times n$ compartmental matrix A is invertible, but also the fact that every entry in $A^{-1} \leq 0$. An interesting method of demonstrating that $A^{-1} \leq 0$ is by means of a Neumann series for A^{-1}. We mimic the approach presented in [4] with some modifications. Let A be non-singular. By way of contradiction, suppose that $a_{jj} = 0$ for some j. Since

$$a_{jj} = - \sum_{i \neq j} a_{ij} - a_{0j}$$

and since both a_{ij} ($i \neq j$) and a_{0j} are nonnegative, we must have $a_{ij} = 0$ for all $i \neq j$. This means that there is no flow of material out of compartment j. Thus the jth compartment is a trap. But this contradicts the invertibility of A. Hence $a_{jj} < 0$ for all j = 1, 2, ..., n. Now since the compartmental matrix A has nonzero diagonal entries, we can form the $n \times n$ diagonal matrix

$$D = \text{diag } \{1/a_{11}, \ 1/a_{22}, \ \dots, \ 1/a_{nn}\}$$

and the error matrix $E \equiv I - AD \equiv [e_{ij}]$.

Exercise 13.3.   Show that the error matrix E has the following properties:

(a) $e_{ij} = -a_{ij}/a_{jj}$ if $i \neq j$;

(b) $e_{ij} = 0$ if $i = j$;

(c) $E \geq 0$;

(d) any column sum of E is less than or equal to unity.

Properties (c) and (d) of Exercise 13.3 clearly show that every entry of E is a small nonnegative number. We may consider the size of E to be a rough estimate of the inaccuracy of D as an approximation to A^{-1}. In particular, if $a_{ij} = 0$ for all $i \neq j$, then $A^{-1} = D$ and E is exactly the zero matrix. Property (a) has the interesting interpretation that e_{ij} is the <u>probability</u> that a particle which is undergoing transition out of compartment j will in fact move to compartment i.

By the construction of the matrix E, it is seen that the <u>1-norm of E</u>, computed by

$$\| E \|_1 \equiv \max_{1 \leq j \leq n} \sum_{i=1}^{n} |e_{ij}| ,$$

satisfies the condition $\| E \|_1 \leq 1$. Since the magnitude of any eigenvalue of E cannot exceed the value of this norm [143], the spectral radius $\rho(E)$ is less than or equal to 1. By way of contradiction, assume $\rho(E) = 1$. Since $E \geq 0$, an appeal to the Perron-Frobenius theory shows that there is a real eigenvalue $\beta = \rho(E) = 1$. Since $\beta I - E = AD$, then

(13.6) $0 = \det(\beta I - E) = \det(AD) = \det A \cdot \det D$.

Clearly D is nonsingular, so we must conclude from (13.6) that $\det A = 0$. But this contradicts the invertibility of A. Hence $\rho(E) < 1$.

Actually, it can be shown (like the proof of Theorem 4 in [4]) that $\rho(E) < 1$ is a necessary as well as sufficient condition that A is invertible (as long as $a_{jj} < 0$ for all j). Hence nonsingularity of A is equivalent to the algebraic problem of demonstrating that every eigenvalue of E has magnitude less than unity. This equivalence is theoretical in nature and not readily applicable to a particular matrix E constructed from A. However, the condition $\rho(E) < 1$ is vitally important for the upcoming discussion as is seen in the next Exercise.

<u>Exercise 13.4.</u>   Show that if a real square matrix B has the property $\rho(B) < 1$, then $(I - B)$ is invertible and $(I - B)^{-1} = I + B + B^2 + B^3 + \dots$.

The equation defining E can be rewritten as

(13.7) $A^{-1} = D(I - E)^{-1}$

where the inverse of $I - E$ exists since A and D are invertible. Application of
Exercise 13.4 to the matrix E yields

(13.8) $\boxed{A^{-1} = D + DE + DE^2 + DE^3 + \ldots}$

from equation (13.7). An expansion of the type in (13.8) is called a <u>Neumann series</u>.
It can be thought of as a regular perturbation expansion of A^{-1} since the series is
in terms of increasing powers of the "small" matrix E, and when E = 0, the inversion
problem has the simple solution $A^{-1} = D$. Each additional term of the sum serves as
a "correction" of one higher order to the preceding terms.

13C. Matrix Inequalities

The series (13.8) can also be developed via the linear iteration scheme

$$X_{p+1} \equiv D + X_p E \qquad\qquad p = 0, 1, 2, \ldots,$$

where X_p, p = 0, 1, 2, ..., represent successive approximations to A^{-1}. If we
select $X_0 \equiv D$, which is the most reasonable choice since D is "close" to A^{-1} in the
sense that the error matrix E is small, then

$$X_1 = D + DE$$

(13.9) $X_2 = D + (D + DE) E = D + DE + DE^2$

$$X_3 = D + (D + DE + DE^2) E = D + DE + DE^2 + DE^3$$

and so on. At this point it is useful to introduce the component-wise partial
ordering $\leq$ (or $\geq$) for any real matrices $G \equiv [g_{ij}]$ and $H \equiv [h_{ij}]$: $G \leq H$ if and only
if $g_{ij} \leq h_{ij}$ for all i, j. Now because $E^j \geq 0$ for all j = 0, 1, 2, ..., and since
$D \leq 0$, then $DE^j \leq 0$ for all j. As a consequence of (13.8), we see that the sequence
$\{X_i\}$ is <u>monotone convergent</u> to A^{-1}:

(13.10) $0 \geq X_0 \geq X_1 \geq X_2 \geq \ldots \geq A^{-1}.$

In (13.10), not only do we have the desired result $0 \geq A^{-1}$, but we also get the more refined information of nonpositive upper bounds on the elements of A^{-1} given by the corresponding elements of the matrices X_0, X_1, X_2, ... In particular, the initial approximation $D \approx A^{-1}$ is improved to $D \geq A^{-1}$.

Exercise 13.5.   Use the inequality $A^{-1} \leq D \leq 0$ to prove the inequality $|\lambda_n| \leq$ min $|a_{ii}|$ (which is already known from Section 12).

The inequality $A^{-1} \leq 0$ can be improved to $A^{-1} < 0$ in certain cases. Let us adopt the notation $(X)_{ij}$ for the (i, j)-element of the matrix X. Then

$$(DE)_{ij} \quad = \sum_{k=1}^{n} (D)_{ik}e_{kj} = e_{ij}/a_{ii}$$

(13.11) $$((DE)E)_{ij} = \sum_{k=1}^{n} (DE)_{ik}e_{kj} = \sum_{k=1}^{n} e_{ik}e_{kj}/a_{ii}$$

$$((DE^2)E)_{ij} = \sum_{k=1}^{n} (DE^2)_{ik}e_{kj} = \sum_{k=1}^{n} \sum_{\ell=1}^{n} e_{i\ell}e_{\ell k}\, e_{kj}/a_{ii}$$

and so on. Thus the (i, j)-element of the second approximation X_1 is directly calculated to be

$$1/a_{ii} \quad \text{if} \quad i = j ; \qquad e_{ij}/a_{ii} \quad \text{if} \quad i \neq j.$$

Due to the fact that $A^{-1} \leq X_1$, the next result obtains. We use the notation $A^{-1} \equiv [a_{ij}^{(-1)}]$.

THEOREM 13.4.   If $a_{ij} \neq 0$ in A, then the corresponding element $a_{ij}^{(-1)} \neq 0$ in A^{-1}. Furthermore,

(13.12) $$a_{ij}^{(-1)} \leq \begin{cases} 1/a_{ii} & \text{if } i = j \\ \\ e_{ij}/a_{ii} & \text{if } i \neq j \end{cases} .$$

Inequality (13.12) indicates that if a particle can move from compartment j to compartment i (i ≠ j) in <u>one</u> transition, then $a_{ij}^{(-1)} < 0$.

<u>Exercise 13.6.</u> Given the compartmental matrix

$$A = \begin{bmatrix} -2 & 1/2 \\ 2 & -2 \end{bmatrix}$$

construct X_0, X_1, X_2, X_3, and confirm the inequalities in (13.10).

<u>Exercise 13.7.</u> Consider the compartmental matrix

$$A = \begin{bmatrix} -4 & 1 & 0 \\ 2 & -5 & 6 \\ 1 & 4 & -6 \end{bmatrix}.$$

Verify the inequalities of (13.12). Compute A^{-1}.

<u>Exercise 13.8.</u> We turn now to the second approximation X_2 of A^{-1}. Prove that the general (i, j)-element in the matrix X_2 is

(13.13) $1/a_{ii} + (1/a_{ii}^2) \sum_{k \neq i} a_{ik} a_{ki}/a_{kk}$

if i = j, and

(13.14) $- a_{ij}/a_{ii} a_{jj} + (1/a_{ii} a_{jj}) \sum_{\substack{k \neq i \\ k \neq j}} a_{ik} a_{kj}/a_{kk}$

whenever i ≠ j.

The inequality $X_2 \geq A^{-1}$ means that $a_{ij}^{(-1)}$ is less than or equal to the appropriate expression in either (13.13) or (13.14). In particular, expression (13.14) will be negative provided there is some k not equal to i or j for which $a_{ik} a_{kj} > 0$. Hence if a particle can move from compartment j to some other compartment k (so $a_{kj} > 0$), and then from compartment k to a different compartment i on the next transition (so $a_{ik} > 0$), then $a_{ij}^{(-1)} < 0$. As an example, consider a <u>mammillary com-</u><u>partmental system</u>. With the mother compartment designated 1, then the associated compartmental matrix is

$$A = \begin{bmatrix} a_{11} & a_{12} & a_{13} & \cdots & a_{1n} \\ a_{21} & a_{22} & 0 & \cdots & 0 \\ a_{31} & 0 & a_{33} & \cdots & 0 \\ \cdot\cdot & \cdot\cdot & \cdot\cdot & \cdots & \cdot \\ a_{n1} & 0 & 0 & \cdots & a_{nn} \end{bmatrix}$$

in which we assume that all a_{i1} and a_{1j} are positive. The connectivity diagram for daughter compartments i and j is in Figure 13.2. It is clear that a particle can move from compartment j to compartment i in two transitions, where the first transition is always to the central compartment. From the above discussion, we must conclude that the inverse of a mammillary matrix must be a matrix all of whose elements are strictly negative, i.e., $A^{-1} < 0$.

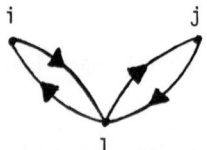

i   j

1

Figure 13.2.

Subsequent inequalities $X_3 \geq A^{-1}$, $X_4 \geq A^{-1}$, ..., and (13.11) lead inductively to the result that if it is possible to pass from some compartment j to any other compartment i in a finite number of transitions, then $a_{ij}^{(-1)} < 0$ in A^{-1}. A special case covered by this statement is that if the compartmental system is strongly connected (or A is an irreducible matrix), then $A^{-1} < 0$. This coincides with a result of Hearon [123, p. 49].

The ideas introduced in this section are closely related to the concept of "M-matrices" [176, p. 108] and nonnegative matrices [76], [143], [153]. A square real invertible matrix $M \equiv [m_{ij}]$ is called an M-matrix if $M^{-1} \geq 0$ and $m_{ij} \leq 0$, $i \neq j$. Thus we see that the negative of an invertible compartmental matrix is an M-matrix.

SECTION 14. MEAN TIMES AND THE INVERSE MATRIX

14A. Mean Residence Times

The entries of the negative inverse of an $n \times n$ compartmental matrix A have an important physical interpretation - namely that of "mean residence times." For this section, we use the open compartmental model given by $\dot{x} = Ax$, $t \geq 0$, where A is invertible. Moreover, we suppose there is a bolus input of tracer at time zero so that $x(0) = x_0 \neq 0$.

Suppose, just for this paragraph, that A is a closed compartmental matrix, that is, all column sums of A are zero. It was observed by Anderson in [9] (also see Section 26) that $Q(h) \equiv (I + hA)^T$ is an $n \times n$ row stochastic matrix for sufficiently small $h > 0$.

Exercise 14.1.   How small must $h > 0$ be in order that $(I + hA)^T$ is a row stochastic matrix?

Let us now return to A being any $n \times n$ open compartmental matrix, i.e., a compartmental matrix where there is at least one j, j = 1, 2, ..., n, such that the excretion constant

$$a_{0j} \equiv - \sum_{i=1}^{n} a_{ij}$$

from compartment j to the outside environment, is positive. From A, we construct a matrix $\bar{A}$, called the closure of A, as follows (we take a simpler approach than in [91]). The outside environment becomes a compartment so that there are now n + 1 compartments in the system. Attach the (n + 1) × 1 vector of zeros on the left-hand side of A, and the 1 × n vector, whose j^{th} entry is a_{0j}, to the top of A to get the (n + 1) × (n + 1) matrix $\bar{A}$.

Exercise 14.2.   Verify that $\bar{A}$ is a closed compartmental matrix.

For a sufficiently small fixed $h > 0$, the (n + 1) × (n + 1) row stochastic matrix $\bar{Q}(h) \equiv (I + h\bar{A})^T$ is the transition matrix for the (n + 1)-state Markov chain

$x(t+h)^T = x(t)^T \bar{Q}(h)$, and has the canonical form

$$\bar{Q}(h) \equiv \left[\begin{array}{c|c} 1 & 0 \\ \hline R(h) & Q(h) \end{array} \right]$$

where $Q(h) = (I + hA)^T$ in terms of the original $n \times n$ compartmental matrix A. This Markov chain is absorbing since the external environment is an absorbing state, and since it is possible to pass from any of the original compartments to the outside (this is because the system has no traps since A is invertible). The matrix

$$M(h) \equiv [I - Q(h)]^{-1} = [I - I - hA^T]^{-1} = \frac{1}{h}(-A^{-1})^T$$

is the <u>fundamental matrix</u> associated with $\bar{Q}(h)$, and its generic element $m_{ij}(h)$ is nonnegative and interpreted as the expected number of times the Markov process is in compartment (state) j given that it began in compartment i and ran until absorbed (i.e., goes to the outside environment) [155, p. 106]. Define $T \equiv - A^{-1} \equiv [\tau_{ij}]$. Then the above equations result in

$$(14.1) \qquad m_{ij}(h) = \tau_{ji}/h \qquad\qquad i, j = 1, 2, \ldots, n.$$

Note that this equation gives an alternative verification that $A^{-1} \leq 0$, as we have seen in Section 13. A Markov chain model is usually set up in <u>unit</u> steps [155]. If we assign $h \equiv 1$ in (14.1), then we arrive at the following result.

THEOREM 14.1. Suppose S is an open compartmental system with no traps, with associated compartmental matrix A. Then the (i, j)-element of $T = - A^{-1}$ is the mean residence time τ_{ij} (with respect to $\bar{Q}$) that a random particle spends in compartment i, having commensed in compartment j at time zero, before exiting the system S.

In the event that A is a singular matrix, the representation of τ_{ij} is different and is studied in [92].

In the <u>scalar</u> case of the compartmental model, we can envision the mean residence time as follows. The system is a one compartment open system as depicted in Figure 14.1, and the model becomes $\dot{x}(t) = a_{11}x(t)$, $x(0) = x_0$, where $a_{11} = - a_{01} < 0$.

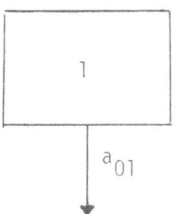

Figure 14.1.

A One Compartment Open System

The curve of the solution $x(t) = x_0 \exp (a_{11}t)$ is shown in Figure 14.2. The mean

time $\tau \equiv \tau_{11} = - a_{11}^{-1}$ according to Theorem 14.1. We think of τ as the average number

(expected value) of time periods until any randomly selected particle leaves the

compartment. For example, if we take a random sample of 1000 particles from the

compartment,. and if for this sample it takes an average of 4 minutes for a particle

to exit the compartment, then $- 1/\tau = - 1/4 = - 0.25$ is a reasonable guess for the

coefficient a_{11} in the model. But also

(14.2) $\tau = - a_{11}^{(-1)} = - x(0)/\dot{x}(0)$

which is the t-axis intercept of the dashed tangent line in Figure 14.2 whose

equation is $y = \dot{x}(0)t + x(0)$. Thus we can interpret τ as that amount of time until

$x = 0$, provided $\dot{x}(t)$ remains constant at its initial value $\dot{x}(0)$.

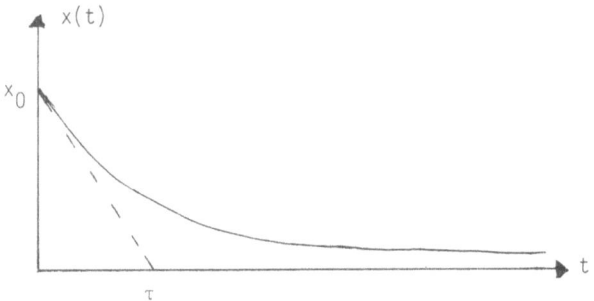

Figure 14.2.

The Mean Residence Time τ

Exercise 14.3. For this 1-compartment model, the _turnover time_ for the compartment is defined as that time τ_1 it takes for the initial load of material x_0 in the compartment to decay by a factor of $1/e$, where e is the natural logarithm base. Is $\tau_1 = \tau_{11}$?

14B. The Compartmental Matrix Exponential

In this subsection, we continue to let A be an $n \times n$ open compartmental matrix which is invertible (unless otherwise specified). Besides

$$(14.3) \qquad Q(t) \equiv (I + tA)^T,$$

we introduce the $n \times n$ compartmental matrix exponential function

$$(14.4) \qquad P(t) \equiv \exp(tA)$$

whose general (i, j)-element $p_{ij}(t)$ has been shown to be the probability that a particle will be in compartment i in time t given that it started in compartment j [91, p. 295]. Even though the elements $p_{ij}(t)$ are probabilities, the matrix P(t) is not necessarily a (column) stochastic matrix. The entries, however, have the properties

$$p_{ij}(t) \geq 0, \qquad \sum_{i=1}^{n} p_{ij}(t) \leq 1$$

for all $t \geq 0$.

Exercise 14.4. [91]. If A is a _closed_ compartmental matrix, show that $\exp(tA)$ is a (column) stochastic matrix for all $t \geq 0$. In particular, if

$$A = \begin{bmatrix} -2 & 1 & 0 \\ 2 & -2 & 0 \\ 0 & 1 & 0 \end{bmatrix} \qquad \begin{aligned} \lambda_1 &= -3.4142 \\ \lambda_2 &= -0.5858 \\ \lambda_3 &= 0. \end{aligned}$$

verify that

$$\exp{(tA)} = \begin{bmatrix} 0.29478 & 0.18541 & 0 \\ 0.37030 & 0.29478 & 0 \\ 0.33492 & 0.51981 & 1 \end{bmatrix}$$

when $t = 1$.

For a <u>closed</u> compartmental matrix A and fixed $t > 0$, the stochastic matrices $P(t) = \exp{(tA)}$ and $Q(t) = (I + tA)^T$ are of course different, with perhaps $P(t)$ <u>less</u> desirable from a modeling viewpoint for the following reason: the directed graph for the matrix $P(t)$ exhibits <u>more</u> connections between compartments than does the connectivity diagram of the compartmental system. This is because, as long as compartment i is reachable from compartment j along <u>some</u> path, the probability $p_{ij}(t) > 0$, even though the fractional transfer coefficient a_{ij} may be zero. On the other hand, the directed graph associated with $Q(t)^T$, $t > 0$, is the <u>same</u> as the connectivity diagram for A. These remarks are demonstrated with the matrices of Exercise 14.4.

Now in an infinitesimal time interval $[t, t+h)$, the mean time a random particle is in compartment i, having been loaded in compartment j, before being excreted, is $p_{ij}(t)h$. As a result, the mean residence time τ_{ij} of a random particle residing in compartment i as a consequence of loading it just in compartment j, would be the (i, j)-element of the $n \times n$ matrix

(14.5) $\int_0^\infty P(t)dt$.

<u>Exercise 14.5.</u>   Demonstrate that

(14.6) $\int_0^\infty P(t)dt = -A^{-1} \equiv T$

(see [143, p. 195]; also see Section 20 on component matrices). Moreover, show that

$$T = \lim_{h \to 0} h[I - P(h)]^{-1}.$$

14C. Further Properties of Mean Residence Time

If flow of material can occur from compartment j to compartment i ($i \neq j$), then $a_{ij} > 0$ and the mean residence time $\tau_{ij} = - a_{ij}^{(-1)}$ should also be positive. This physically obvious fact is confirmed in the mathematical model via inequality (13.12) from which we see that

$$\tau_{ij} \geq a_{ij} / a_{ii} a_{jj} > 0.$$

Also from (13.12),

$$(14.7) \qquad \tau_{ii} \geq - 1/a_{ii} \equiv \tau_i .$$

This generalizes (14.2) to multicompartmental systems. That strict inequality can hold in (14.7) is due to the fact that in a multicompartmental system in which flow can occur between at least two distinct compartments, the particles may return to compartment i many times. If compartment i is not reachable from any other compartment, then particles will exit from compartment i only once thus giving $\tau_{ii} = - 1/a_{ii} = \tau_i$, the turnover time for compartment i.

There is certainly one situation in which the mean times are easily described and it again is a generalization of (14.2). Suppose S is a single exit compartmental system with excretion only from the mth compartment.

Exercise 14.6. [4]. Prove that if the SEC matrix A is invertible, then each entry in the mth row of $- A^{-1}$ is $1/a_{0m}$. Then the mean times $\tau_{mj} = 1/a_{0m}$ for all compartments $j = 1, 2, \ldots, n$.

The physical interpretation of this result is that since A is invertible, there are no traps in S, and so all particles must eventually enter compartment m to exit the system, regardless of where they were loaded into the system. Moreover, their time in the exit compartment is inversely proportional to the excretion rate from that compartment.

Consider the SEC example matrix

$$A = \begin{bmatrix} -4 & 1 & 2 \\ 2 & -5 & 4 \\ 1 & 4 & -6 \end{bmatrix}$$

for which

$$T = - A^{-1} = \frac{1}{14} \begin{bmatrix} 14 & 14 & 14 \\ 16 & 22 & 20 \\ 13 & 17 & 18 \end{bmatrix} .$$

In this case, $m = 1$, $a_{01} = 1$, and $\tau_{1j} = 1$ for $j = 1, 2, 3$.

Now let us assume that the ith compartment of the model

$$\dot{x} = Ax, \quad x(0) = x_0 \equiv [x_{10}, x_{20}, \ldots, x_{n0}]^T$$

is sampled over time so that we can assume that $x_i(t)$ is known (at least at discrete times). Consider the sum

$$\sum_{j=1}^{n} \tau_{ij} x_{j0}$$

which is just the ith component of Tx_0. Then $\dot{x} = Ax$ implies that $T\dot{x} = -x$, or

$$T \int_0^\infty \dot{x}(t) dt = - \int_0^\infty x(t) dt.$$

Since $x(\infty) = 0$, then we must have, upon integrating,

(14.8)
$$\sum_{j=1}^{n} \tau_{ij} x_{j0} = (Tx_0)_i = \int_0^\infty x_i(t) dt.$$

Thus in particular, if we load only the mth compartment with an amount $x_{m0} = D$, then

(14.9)
$$\boxed{\tau_{im} = \int_0^\infty x_i(t) dt / D.}$$

This formula provides a method of directly estimating the mean time τ_{im} which does not require knowledge of the entries of A. Note that is x_i is observed only at discrete times, then the integral in (14.9) can be evaluated by numerical quadrature, thereby avoiding a curve-fitting problem for x_i.

As a simple example, let the mathematical model for tracer kinetics in a two compartment system be

$$\dot{x} = \begin{bmatrix} -2 & 1 \\ 1 & -2 \end{bmatrix} x, \qquad x_0 = \begin{bmatrix} 1 \\ 0 \end{bmatrix} \equiv \begin{bmatrix} D \\ 0 \end{bmatrix}.$$

Then the unique solution of these equations is

$$x_1(t) = \exp(-3t)/2 + \exp(-t)/2$$

$$x_2(t) = -\exp(-3t)/2 + \exp(-t)/2.$$

Assume that the second compartment is observed. Then

$$\tau_{21} = \int_0^\infty x_2(t)dt = 1/3$$

according to formula (14.9).

Exercise 14.7. Suppose that a compartmental system is single exit through compartment i, and that compartment i is initially loaded with a unit amount of tracer and is also sampled over time. By summing the columns of $\dot{x} = Ax$, show that (14.9) leads to $\tau_{ii} = 1/a_{0i}$, which agrees with the result in Exercise 14.6.

THEOREM 14.2. Let A be a nonsingular compartmental matrix. Then the maximal element of the nonnegative matrix $T = -A^{-1}$ lies on its diagonal.

Proof. We only need to show that

$$r \equiv \max_{i,k} \tau_{ik} \leq \max_i \tau_{ii} \equiv s.$$

Let j be an arbitrary but fixed integer in {1, 2, ..., n}. For any $i \neq j$, the equation $-I = TA$ implies

(14.10) $0 = \tau_{ii}a_{ij} + \sum_{k \neq i} \tau_{ik}a_{kj}$

$\leq \tau_{ii}a_{ij} + (\max_k \tau_{ik}) \sum_{k \neq i} a_{kj}$.

Summing both sides of (14.10) over all $i \neq j$,

(14.11) $0 \leq \sum_{i \neq j} \tau_{ii}a_{ij} + \sum_{i \neq j} \left[(\max_k \tau_{ik}) \sum_{k \neq i} a_{kj} \right]$

$\leq s \sum_{i \neq j} a_{ij} + r \sum_{i \neq j} \sum_{k \neq i} a_{kj}$.

The double sum in (14.11) is equal to

(14.12) $(n-1) \sum_{k \neq j} a_{kj} + (n-1)a_{jj} - \sum_{k \neq j} a_{kj} = (n-2) \sum_{k \neq j} a_{kj} - (n-1)|a_{jj}|$.

In (14.11) and (14.12), replace the jth column sum (skipping the diagonal element) by the larger quantity $|a_{jj}|$. Then (14.11), with the substitution of (14.12) yields

$0 \leq s|a_{jj}| + r[(n-2)|a_{jj}| - (n-1)|a_{jj}|] = (s-r)|a_{jj}|$.

Since $|a_{jj}| > 0$, then $s - r \geq 0$.

▨

The related inequality $\tau_{ij} \leq \tau_{ii}$ is also true for all i, $j = 1, 2, \ldots, n$ [6, p. 68]. This is also clear from the physical standpoint because of the interpretation of the τs as mean times. Equality $\tau_{ij} = \tau_{ii}$ can occur for all j as we have seen in Exercise 14.6.

14D. System Mean Residence Time

Now let us turn our attention to the <u>sum</u> of mean residence times for each compartment:

(14.13) $\sum_{i=1}^{n} \tau_{im}$.

We might call this the <u>mean residence time of a particle in the entire system</u>, given that it was loaded in compartment m. Hearon [120] has studied this system mean residence time. For small h > 0, we observe that

$$\sum_{i=1}^{n} a_{0i} x_i(t) h$$

is approximately the number of tracer particles eliminated from the system in the time interval $[t, t+h]$. Thus

$$\int_0^\infty \sum_{i=1}^{n} a_{0i} x_i(t) dt$$

is the total number of tracer particles eliminated from the compartmental system over all time. Since the system has no traps (A is invertible), this total number of tracer particles eliminated is exactly the total amount

$$D \equiv \sum_{i=1}^{n} x_{i0}$$

of tracer injected into the system initially. Define

$$\Psi(t) \equiv \sum_{j=1}^{n} [a_{0j} x_j(t)]/D.$$

From the above discussion, it follows that

$$\Psi(t) \geq 0, \quad t \geq 0; \qquad \int_0^\infty \Psi(t) dt = 1.$$

Because of this, Ψ is the <u>distribution function of exit times</u>. Furthermore, since all tagged particles were present in the system at $t = 0$, Ψ also is the distribution of residence times for particles in the system.

<u>Exercise 14.8.</u>   Prove that

$$(14.14) \qquad \sum_{j=1}^{n} \dot{x}_j(t) = -\Psi(t)D$$

for all $t \geq 0$.

Let us now define the moments of the distribution Ψ:

$$\mu_r \equiv \int_0^\infty t^r \Psi(t)dt \qquad r = 0, 1, 2, \ldots .$$

In particular, it is calculated that

$$\mu_1 = \int_0^\infty t\Psi(t)dt = - \int_0^\infty t \sum_{i=1}^n \dot{x}_i(t)/Ddt$$

$$= (- 1/D) \sum_{i=1}^n \int_0^\infty t\dot{x}_i(t)dt .$$

The last integral is integrated by parts, in which we take

$$tx_i(t) \to 0, \qquad t \to \infty$$

for each compartment i.   Hence

$$\mu_1 = (1/D) \sum_{i=1}^n \int_0^\infty x_i(t)dt = (1/D) \sum_{i=1}^n \sum_{j=1}^n \tau_{ij}x_{j0}$$

from (14.8).   As a consequence, under the commonly encountered initial condition that the mth compartment is the only one loaded, with a bolus amount $x_{m0} = D$ of tracer, then

$$\mu_1 = (1/D) \sum_{i=1}^n \tau_{im}x_{m0} = \sum_{i=1}^n \tau_{im} .$$

Hence we see that Hearon's distribution of residence times for particles in the system is consistent with our mean residence time of a particle in the system, as presented in (14.13).

SECTION 15. SOLUTION OF THE STEADY STATE PROBLEM FOR SEC SYSTEMS

15A. The Tracer Steady State Problem

At this time let us consider the model

$$\dot{x}(t) = A\, x(t) + b, \qquad x(0) = x_0,$$

for the tracer kinetics in an n-compartment system in the case of a constant
infusion of tracer over a long time period. It is assumed that the system is open,
that b is a nonnegative constant tracer input vector over time, and let us suppose
that the system contains no traps so that A is invertible. We have already seen
that under these conditions, as $t \to \infty$, there is a unique steady state vector x_e
given by $Ax_e + b = \dot{x}_e = 0$, or

(15.1) $x_e = A^{-1}(-b)$.

The solution of this linear algebraic system $Ax_e = -b$ is normally straightforward
by direct methods of solution such as Gaussian elimination. If there is a small
degree of connectivity in the system in the sense that many $a_{ij} = 0$, that is, if
the system matrix A is exceptionally sparse, then an iterative method of solution
may be more efficient and hence preferable to a direct method.

15B. Ill-Conditioned SEC Systems

There is one particular case of this problem where difficulties can arise and
where an iterative technique comes to the rescue. That case is for certain types
of SEC systems. In about 15% - 20% of the SEC cases we have faced, the inversion
problem (15.1) was sufficiently "ill-conditioned" that when using direct methods
such as LEQT1F in the IMSL package [131], the verdict was returned that A was
"computationally singular." The actual problem at hand was usually that a very
small pivot element had emerged in the process of executing the direct method and
consequently the algorithm for finding x_e automatically terminated.

The ill-conditioned problem for the solution of the linear algebraic system
means that slight changes in the entries in A (which are usually derived from

experimental data and hence can contain significant errors) may precipitate the undesirable situation of very large changes in the values of the solution vector x_e. An example of the severity of conditioning in this inversion problem is shown with the SEC matrix

$$(15.2) \qquad A = \begin{bmatrix} -0.8 - a_{01} & 0.2 & 0.4 \\ 0.7 & -0.7 & 0.7 \\ 0.1 & 0.5 & -1.1 \end{bmatrix}.$$

With $a_{01} = 5.0 \times 10^{-3}$, the inverse matrix is calculated to be

$$A^{-1} \doteq \begin{bmatrix} -200. & -200. & -200. \\ -400. & -403. & -402. \\ -200. & -201. & -202. \end{bmatrix}.$$

However, let us suppose that the excretion rate a_{01} is in error through measurement by 1.0×10^{-3}, so that a_{01} is really 4.0×10^{-3}. With this true excretion rate constant, the inverse matrix is reported as

$$A^{-1} \doteq \begin{bmatrix} -250. & -250. & -250. \\ -500. & -503. & -502. \\ -250. & -251. & -252. \end{bmatrix}.$$

Thus we observe that a change of one unit in just one element of A yields as much as a change of 100,000 units in several entries of A^{-1}!

In actual computing situations where such severe problems are encountered, even other signs of ill-conditioning in the SEC matrix A may present themselves. Observe that

(a) det A approaches zero as $a_{0m} \rightarrow 0$, and hence the matrix approaches singularity;

(b) the elements of A^{-1} (or any approximate inverse) may be quite large;

(c) the matrix A has a large "condition number."

Let us discuss these conditions one by one. If a_{0m} is set equal to zero, then by Exercise 12.1, we see that every column sum of A is zero, and thus det A = 0. It follows that if $a_{0m} \approx 0$, then det A is approximately zero by some measure; certainly det A → 0 as a_{0m} → 0.

With regard to (b), one gauge of how large the elements are in A^{-1} is given in Exercise 14.3: Assuming the SEC matrix A is invertible, then each entry in the mth row of A^{-1} is $- 1/a_{0m}$. Thus if a_{0m} is quite small, then certainly the mth row elements in A^{-1} will be large negative numbers.

In dealing with the general linear algebraic system Bx = y, the <u>condition number</u> of the invertible coefficient matrix $B \equiv [b_{ij}]$ is usually defined by

$$\text{cond } (B) \equiv \| B \| \ \| B^{-1} \|$$

after an appropriate matrix norm $\| \cdot \|$ is selected. This condition number is always greater than or equal to one. If it is quite large relative to the magnitudes of the entries in the matrix B, then ill-conditioning of the linear system Bx = y is to be expected. Note, however, that a large cond(B) does not necessarily imply the problem is ill-conditioned. Normally we do not know B^{-1} and hence $\| B^{-1} \|$ cannot be computed. However, in the case of a SEC matrix A, this difficulty can be circumnavigated in that a lower bound for cond(A) can be directly computed from knowledge of the elements of A.

<u>Exercise 15.1.</u> Since we know something of the row structure of A^{-1}, let us select the ∞-norm [143] which, for a general n × n matrix B, is computed by

$$\| B \|_{\infty} = \max_{1 \le i \le n} \sum_{j=1}^{n} |b_{ij}| .$$

Using this norm, prove that cond$(A) \ge n \| A \|_{\infty} / a_{0m}$, where A is a SEC matrix with exit from the mth compartment.

To illustrate this result, consider the SEC matrix given in (15.2) with $a_{01} = 5.0 \times 10^{-3}$. We have $\|A\|_{\infty} = 2.1$ so that

$$\text{cond}\,(A) \geq 3(2.1)/5.0 \times 10^{-3} = 1260,$$

a strong indicator of ill-conditioning in this particular matrix inversion problem.

15C. An Iterative Procedure for SEC Systems

As has been indicated, when A is an SEC matrix, direct methods of elimination may fail in computing the solution x of $Ax = y$ for some prescribed $n \times 1$ vector y. In the present section this problem is avoided by developing a numerically stable iterative procedure for computing $x = A^{-1}y$. As will be shown, the method works because it is constantly self-correcting. Other methods such as rescaling, singular value decomposition, and Tychonov regularization may also be tried [170], [108], [22].

For the iterative process to be developed, the central concept used is an "approximate inverse" - like that in Section 13 used in proving the inverse of a compartmental matrix is nonpositive. The difference is that we utilize a left approximate inverse "J" for a given SEC matrix A instead of a right approximate inverse D for a general compartmental matrix as discussed in Section 13. Our approach is based on reference [4]; see [234], [176] for information on general iterative methods.

Given a nonsingular $n \times n$ SEC matrix A, suppose we can construct an invertible $n \times n$ approximate inverse J of A such that the error matrix

$$E \equiv [e_{ij}] \equiv I - JA$$

has the property that its spectral radius $\rho(E) < 1$ [176]. The solution vector $x = A^{-1}y$ is a fixed point of the equation $x = Ex + Jy$, and hence the linear iterative scheme

$$x^{(p+1)} \equiv E\,x^{(p)} + Jy \qquad\qquad p = 0, 1, 2, \ldots$$

is suggested. We hope vectors $x^{(1)}$, $x^{(2)}$, ..., will converge to the solution vector x. This convergence does take place since the condition $\rho(E) < 1$ is met.

Let us pause a moment and ask why we might want to use a left inverse form JA rather than a right inverse combination AJ. Let $x = A^{-1}y$ be the exact solution of the problem and define $x^{(0)} \equiv Jy$, a first order approximate solution. Then

$$x - x^{(0)} = (A^{-1} - J)y = (I - JA)A^{-1}y = (I - JA)x,$$

so that the <u>relative error</u>

$$\| x - x^{(0)} \| / \| x \| \leq \| I - JA \|$$

for whatever appropriate norm is chosen. Hence the relative error in the estimate $x^{(0)}$ is small provided J is a true left approximate inverse of A, that is, we would <u>want the condition $\| I - JA \| < 1$</u>. On the other hand,

$$x - x^{(0)} = (A^{-1} - J)y = A^{-1}(I - AJ)y$$

shows that smallness of $\| x - x^{(0)} \|$ is <u>not</u> necessarily guaranteed even if I - AJ is small in norm.

A left approximate inverse that always works is constructed as follows for a given $n \times n$ SEC matrix $A \equiv [a_{ij}]$ with exit compartment m. Because J should look as much like A^{-1} as possible, and since every entry in the mth row of A^{-1} is $- 1/a_{0m}$, we let every entry in the mth row of J be $- 1/a_{0m}$. In Section 13, diagonal entries of $1/a_{ii}$ were used in the right approximate inverse D. Here we modify that choice a bit. Instead, let the diagonal entries of J (in all but the (m,m)-position) be $- 1/c$, where $c \equiv \max|a_{ii}|$, $i = 1, 2, ..., n$. (The reason for this particular selection of diagonal element is that it will later enable us to show that $\| E \|_1 \leq 1$.). All remaining entries in J will be zero. As an elementary example, consider the SEC matrix

$$A = \begin{bmatrix} -4 & 1 & 2 \\ 2 & -5 & 4 \\ 1 & 4 & -6 \end{bmatrix}$$

in which $m = 1$, $a_{0m} = 1$, and $c = 6$. Then according to our rule, the J associated with this A is

$$J = \begin{bmatrix} -1 & -1 & -1 \\ 0 & -1/6 & 0 \\ 0 & 0 & -1/6 \end{bmatrix}.$$

<u>Exercise 15.2.</u>   Show that the J matrix is invertible and discover the general form of J^{-1}. Verify that J^{-1} is itself an SEC matrix.

<u>Exercise 15.3.</u>   Prove that every entry of E is nonnegative but small enough that each column sum in E is less than or equal to unity. What can be said about the mth row of E?

<u>Exercise 15.4.</u>   If the SEC matrix A is nonsingular, show that the spectral radius $\rho(E) < 1$.

Now consider the results of our iterative scheme:

$$x^{(1)} \equiv Jy + Ex^{(0)} = Jy + EJy$$

$$x^{(2)} \equiv Jy + Ex^{(1)} = Jy + EJy + E^2Jy$$

$$x^{(3)} \equiv Jy + Ex^{(2)} = Jy + EJy + E^2Jy + E^3Jy$$

and so forth. Thus

$$x \equiv \lim_{p \to \infty} x^{(p)} = \left(\sum_{p=0}^{\infty} E^p \right) Jy \qquad (E^0 \equiv I).$$

As in Exercise 13.4, this series converges because we now know $\rho(E) < 1$. Furthermore, it follows that

$$x = (I - E)^{-1}Jy = (JA)^{-1}Jy = A^{-1}y,$$

the unique solution of our system $Ax = y$.

More can be said due to the physical system we are modeling. Since the equation $Ax = y$ represents the steady state tracer condition given in $Ax_e = -b$, and since $b \geq 0$, then $y \equiv -b \leq 0$. As a consequence, the inequalities $E \geq 0$, $J \leq 0$, and $x^{(0)} \equiv Jy \geq 0$ combine to imply that the successive approximations are <u>monotone</u> <u>increasing</u> in their convergence to x:

$$0 \leq x^{(0)} \leq x^{(1)} \leq x^{(2)} \leq \ldots \leq x.$$

When the iterative scheme is written in component form, which should be used for actual computation on most machines, we have

$$x_i^{(p+1)} = \sum_{j=1}^{n} e_{ij} x_j^{(p)} + x_i^{(0)} \qquad\qquad i = 1, 2, \ldots, n.$$

The number of these equations can be reduced by one since the exact solution for the mth component of $x = A^{-1}y$ is known:

$$x_m = \sum_{j=1}^{n} a_{mj}^{(-1)} y_j = (- 1/a_{0m}) \sum_{j=1}^{n} y_j .$$

Hence, in the algorithm, every $x_m^{(p)}$ can be replaced by x_m.

<u>Exercise 15.5.</u>   Using the algorithm developed in this Section, compute the solution of $Ax = y$ where A is given in (15.2) with $a_{01} = 5.0 \times 10^{-3}$, and $y = (- 1, 0, 0)^T$. What is your stopping criterion?

15D.   Updating the Algorithm

There is a very practical inefficiency with our algorithm for computing x. The scheme requires us to retain all components of $x^{(p)}$ until the calculation of $x^{(p+1)}$ is finished. We can cut computer storage by splitting the scheme,

$$x_i^{(p+1)} = \sum_{j=1}^{i-1} e_{ij} x_j^{(p)} + \sum_{j=i}^{n} e_{ij} x_j^{(p)} + x_i^{(0)}$$

$i = 1, 2, \ldots, n$, and in the first sum replace $x_j^{(p)}$ by $x_j^{(p+1)}$ (assuming the computations of the components of the vector $x^{(p)}$ are done sequentially). Thus we start using each component of the new $x^{(p+1)}$ as soon as it is calculated. The old components of $x^{(p)}$ may be destroyed as $x^{(p+1)}$ is being created. Because this <u>updating procedure</u> makes use of more recent information, it will probably converge, and converge faster than the previous algorithm.

To analyze the question of acceleration of convergence, we move to a matrix format. Let L denote the strictly lower triangular part of the matrix E, and let U be the upper triangular part of E including its diagonal. Then the computation of the vectors in the updating process is given by

$$x^{(p+1)} = L \, x^{(p+1)} + U \, x^{(p)} + x^{(0)} \, ,$$

or equivalently by

$$x^{(p+1)} = (I - L)^{-1} \, U \, x^{(p)} + (I - L)^{-1} \, Jy \, .$$

<u>Exercise 15.6.</u>   Why is the matrix I - L nonsingular?

Note that the new computing method is of the general iterative form

$$x^{(p+1)} = Q \, x^{(p)} + d$$

where

$$Q \equiv (I - L)^{-1} \, U, \qquad d \equiv (I - L)^{-1} \, Jy.$$

In order that this sequence of iterates converge, it is sufficient that $\rho(Q) < 1$. This has been shown to be the case [4].

<u>Exercise 15.7.</u>   Show that the sequence $\{x^{(p)}\}$ of iterates generated via the updating scheme is monotone increasing.

<u>Exercise 15.8.</u>   Verify that the updating procedure has at least as fast a rate of convergence as the original algorithm. This can be done as follows. For the general linear iterative vector scheme

$$x^{(p+1)} \equiv M \, x^{(p)} + g \qquad\qquad p = 0, 1, 2, \ldots \, ,$$

where M is a square matrix, the smaller $\rho(M)$ is, the faster the rate of convergence of the scheme [176]. Hence prove that $\rho(Q) \leq \rho(E)$, where Q is defined above as the coefficient matrix in the updating scheme, and E is the matrix in the original iterative method.

Exercise 15.9.   Using the updating process, solve the system treated in Exercise 15.5.   How does the rate of convergence of the updating scheme compare with that of the original method?

SECTION 16.   STRUCTURAL IDENTIFICATION OF THE MODEL

16A.   The System (A, B, C)

A central problem in linear compartmental studies is discussed in this section. The question is whether or not the system parameters a_{ij} in the model $\dot{x} = Ax + b$, $A \equiv [a_{ij}]$, can be uniquely determined from experimental observation or measurement of certain components of the solution vector x.   The closely related problem of getting actual numerical estimates of the a_{ij}s will be considered later in Section 19.

Usually there are a number of plausible system matrices that fit the observed components of the model.   This number can be reduced by biological or modeling constraints on the elements of the matrices.   An important question is when can A be made unique by the imposition of such constraints.   We will show that there are classes of compartmental models for which the model matrix A can be uniquely determined from experimental observations.

To systematically treat the identification problem, we will slightly modify our model

(16.1) $\dot{x}(t) = A \, x(t) + b(t)$, $t \geq 0$

$$x(0) = x_0.$$

In most applications of compartmental modeling and tracer studies, the tracer input vector b(t) in (16.1) is derived from a single, or maybe a few, controlled injections into the system.   For purposes of analysis, it is desirable to set up a mechanism for the structural connection between the sources of input and the resulting vector b(t).   To explicitly display this structural relation in the model (16.1), we rewrite the system to account for this additional structure.   The ith equation in (16.1) is

(16.2) $\dot{x}_i = \sum\limits_{j=1}^{n} a_{ij} x_j + b_i.$

Expand the forcing term $b_i(t)$ in (16.2) as

$$\sum_{k=1}^{q} b_{ik} u_k(t), \qquad i = 1, 2, \ldots, n,$$

where the functions $u_1(t)$, $u_2(t)$, $\ldots$, $u_q(t)$ are called the <u>input functions</u> of the system. If we let

$$u(t) \equiv [u_1(t) \quad u_2(t) \quad \ldots \quad u_q(t)]^T$$

be the $q \times 1$ input vector and define the $n \times q$ matrix $B \equiv [b_{ik}]$, then in matrix-vector form, system (16.1) becomes

(16.3) $\qquad \dot{x}(t) = Ax(t) + Bu(t), \qquad\qquad t \geq 0$

$$\qquad\qquad x(0) = x_0.$$

Here B is referred to as the <u>input distribution matrix</u> - for it indicates how the multiple inputs $u_i(t)$ are distributed throughout the system: entry b_{ik} is positive if input $u_k(t)$ enters compartment i, and zero otherwise. It is not unusual for $q = 1$, wherein we have a single control function $u(t)$; in such a case B is just an $n \times 1$ column vector. For instance, if there is continuous input only into compartment one, then B could be selected as $[1 \quad 0 \quad \ldots \quad 0]^T$.

To (16.3) we attach the <u>output function</u> which is the vector

(16.4) $\qquad y(t) \equiv Cx(t)$

where C is called the <u>output connection matrix</u> and is a $p \times n$ matrix consisting of constant nonnegative entries c_{ij}. This matrix indicates the paths from compartments to sampling devices in that c_{ij} is positive if compartment j influences output function component y_i; otherwise $c_{ij} \equiv 0$. For example, if the first and third components of x are observed over time in a 5-compartment model, then we could choose

$$C \equiv \begin{bmatrix} 1 & 0 & 0 & 0 & 0 \\ 0 & 0 & 1 & 0 & 0 \end{bmatrix}$$

so that the system output is $y = Cx = [x_1 \quad x_3]^T$.

It is standard to take $x_0 \equiv 0$ in (16.3) and instead incorporate the effect of initial input in the model through the distribution combination Bu. Writing the model complete with input and output structure gives

(16.5)
$$
\begin{array}{l}
\dot{x}(t) = Ax(t) + Bu(t), \qquad t \geq 0 \\[2ex]
x(0) = 0, \\[2ex]
y(t) = Cx(t) ,
\end{array}
$$

and (16.5) will be referred to as the system (A, B, C).

16B. The Structural Identification Problem

The typical tracer experiment is to select an input u and a distribution B, decide what compartments or combination of compartments can be sampled (i.e., specify C), and then record the output function $y(t) = Cx(t)$, usually at discrete times. By this experiment we have a collection of "input-output" data from which to determine if possible the entries in A.

The structural identification problem can now be stated as follows: Does there exist a method by which the unknown parameters in A, B, C can be uniquely determined from the selected experimental setup and its resulting input-output data? The related parameter estimation problem is carrying out the process of actually estimating the numerical values of these unknown parameters from the input-output data. The most common criterion for establishing the parameter estimation and identification is the unique minimization of a sum-of-squares function

$$
S \equiv \sum_{i=1}^{m} \| y(t_i) - d_i \|^2
$$

where the d_i are data vectors, or equivalently, the unique solution of a nonlinear system of equations. Additional information on the identification of linear dynamic systems can be found in [36], [128], [15], [62], [116], [171], [78].

Bellman and Astrom [24] are generally credited with bringing the structural identification problem into the framework of modern control theory by writing the model as in (16.5) and noting that the input-output relation is given via

(16.6)
$$y(t) = Cx(t) = C \int_0^t [\exp(t - \tau)A] Bu(\tau) d\tau$$
$$\equiv [C \exp(tA) B] * u(t),$$

in which * denotes the <u>convolution operation</u>. Hence the model parameters enter the input-output relation only in the <u>impulse response function</u>

(16.7) $\phi(t) \equiv C \exp(tA) B,$ $t \geq 0$,

a matrix of dimensions $p \times q$, or equivalently, in the <u>transfer function</u>

(16.8) $\hat{\phi}(s) = C[sI - A]^{-1} B,$

a $p \times q$ matrix which is the Laplace transformation of ϕ. We can thus conclude from (16.6) - (16.8) that <u>all</u> identifiable system parameters must be identified through either ϕ or $\hat{\phi}$. What makes the structural identification problem difficult is that either ϕ or $\hat{\phi}$ are matrices whose entries are <u>nonlinear</u> expressions in the elements a_{ij}, b_{ij}, and c_{ij}. The system (A, B, C) is <u>identifiable</u> when this nonlinear system of equations has a <u>unique</u> solution. Usually the elements of B and C are known, and so the problem reduces to identifying A.

Taking the Laplace transformation of (16.6) yields

(16.9) $\hat{y} = \hat{\phi} \hat{u}$.

If both y and u are known functions for all time, then $\hat{\phi}$ is known via (16.9), and then inversion methods can give an analytic expression for ϕ. Unfortunately, in real experiments y(t) (and sometimes u(t)) is known only empirically and hence we must go through some type of curve-fitting process to get a known analytical representation of y (see Section 23).

A very common tracer experiment involves the <u>bolus input hypothesis</u> which means that a single known input u(t) of D units of tracer is applied to the system instantaneously at time zero. We represent this in our model by $u(t) = D \delta(t)$, where δ is the <u>Dirac delta function</u> which has the property that $\phi * \delta = \phi$ for appropriate functions ϕ [111, p. 46]. For this input u = Dδ, the corresponding system output is the p × 1 vector

(16.10)   y = ϕ * u = D(ϕ * δ) = Dϕ.

Thus a known constant multiple of ϕ is what is actually being observed in the bolus input experiment.

We will assume for the remainder of this section that B and C are known matrices and that ϕ (or $\hat{\phi}$) is a known function from tracer sampling, and concern ourselves with the question of when can the unknown model parameters a_{ij} of A be uniquely determined from knowledge of the impulse response parameters in ϕ (or $\hat{\phi}$).

16C. A Simple Identification Example

As an example let us consider the general two-compartment model (Figure 16.1) in which a single bolus input $\delta(t)$ is made to the first compartment (so B $\equiv$ [1   0]^T)

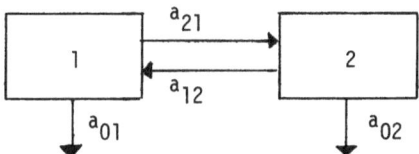

Figure 16.1.

and only the first compartment is sampled in time (thus C = [1   0]). The system matrix is

$$A = \begin{bmatrix} a_{11} & a_{12} \\ a_{21} & a_{22} \end{bmatrix} \qquad \begin{array}{l} a_{11} = -a_{21} - a_{01} \\[2ex] a_{22} = -a_{12} - a_{02} \end{array} .$$

Exercise 16.1.   Compute $\hat{\phi}(s)$ and show that it is

$$(s + \gamma_1) / (s^2 + \gamma_2 s + \gamma_3)$$

where

$$\gamma_1 = a_{12} + a_{02}$$

(16.11) $$\gamma_2 = a_{21} + a_{01} + a_{12} + a_{02}$$

$$\gamma_3 = a_{01}a_{12} + a_{01}a_{02} + a_{21}a_{02} \ .$$

We are assuming $\hat{\phi}$ is known, that is, γ_1, γ_2, and γ_3 are known constant impulse response parameters. From the above calculations, any set $\{a_{ij}\}$ of model parameters for which equations (16.11) hold yield the same input-output information. Since the three nonlinear equations (16.11) contain four unknowns a_{12}, a_{01}, a_{21}, and a_{02}, it is clear that the a_{ij} cannot be uniquely determined from this experiment, i.e., A is not identifiable.

Suppose, however, it is possible to redesign the experiment by clamping off the exit from the second compartment. Thus the constraint $a_{02} = 0$ is imposed. Then it can be shown that the remaining parameters a_{12}, a_{01}, and a_{21} are uniquely determined from equations (16.11) and hence under these conditions A is identifiable.

Exercise 16.2.   Verify that if either a_{01} and/or a_{02} is zero, then the above matrix A is identifiable.

Note that this result confirms the identifiability of the closed system matrix used in the kinetics of the uptake of potassium by the red blood cells which was studied in Section 8 (with a relabeling of compartments), and also the identifiability of a number of other models such as the two compartment model used to facilitate the diagnosis of renal disease [29], or the two compartment model for the study of intravenously injected creatinine and its clearance [205].

16D. Realizations of Impulse Response Functions

For an observed function $\phi(t)$, $t \geq 0$, it is essential that we know a priori that there exists a system (A, B, C) that has ϕ as its impulse response function. If there is such a compartmental system, it is called a realization of ϕ. For instance, suppose through data-gathering we observe the function

$$\phi(t) = - \exp(- 2t)/2 + 1/2.$$

Then the system (A, B, C) where

$$A = \begin{bmatrix} -1 & 1 \\ 1 & -1 \end{bmatrix} \qquad B = \begin{bmatrix} 1 \\ 0 \end{bmatrix} \qquad C = \begin{bmatrix} 0 & 1 \end{bmatrix}$$

is a realization of ϕ since

$$\exp(tA) = \frac{1}{2} \begin{bmatrix} \exp(-2t) + 1 & -\exp(-2t) + 1 \\ -\exp(-2t) + 1 & \exp(-2t) + 1 \end{bmatrix}$$

and therefore

$$C \exp(tA) B = -\exp(-2t)/2 + 1/2 = \phi(t).$$

Exercise 16.3. Realizations of an impulse response function are not unique. Given that (A, B, C) is a realization of ϕ, one way of constructing another realization of ϕ is as follows [89]. Let H be any invertible matrix having the same dimension as A and define

(16.12) $A_1 \equiv H^{-1} A H$ $B_1 \equiv H^{-1} B$ $C_1 \equiv C H$.

Prove that (A_1, B_1, C_1) is also a realization of ϕ.

16E. Impulse Response Function Structure

In this paragraph we get at a more explicit description of the impulse response function ϕ. Suppose (A, B, C) is a realization of ϕ. In practice it is commonplace to have distinct real eigenvalues λ_i of A. Let us assume this condition holds for the moment and define, as before, $S \equiv [s_{ij}]$ to be the $n \times n$ matrix with the corresponding eigenvectors as its columns. Since these eigenvectors are linearly independent,

$$S^{-1} \equiv \begin{bmatrix} s_{ij}^{(-1)} \end{bmatrix}$$

exists. Thus A is diagonalizable, $S^{-1} A S = \Lambda$, where Λ is the $n \times n$ diagonal matrix of eigenvalues $\{\lambda_1 \ \lambda_2 \ \ldots \ \lambda_n\}$. We have also seen that

$$\exp(tA) = S \exp(t\Lambda) \ S^{-1}.$$

Now let $H \equiv S$ in (16.12). Then $(\Lambda, \ S^{-1}B, \ CS)$ is also a realization of ϕ:

(16.13) $CS \exp(t\Lambda) \ S^{-1}B = C \exp(tA) \ B = \phi(t).$

In (16.13) let

(16.14) $K \equiv CS \equiv [k_{ij}] \qquad L \equiv S^{-1}B \equiv [\ell_{ij}].$

Then upon direct expansion of (16.13), the general (i, j)-entry of the $p \times q$ matrix $\phi(t)$ is

(16.15) $\displaystyle\sum_{m=1}^{n} k_{im} \ \ell_{mj} \ \exp(\lambda_m t).$

Besides having distinct real eigenvalues for A, it is also typical in tracer problems to have input to just one compartment and to have only one compartment accessible to measurement. Let e_k denote the kth column (vector) of the $n \times n$ identity matrix. If only the jth compartment is injected with tracer, we can select $B \equiv e_j$. If tracer sampling is done only on the ith compartment, then $C \equiv e_i^T$. Under these conditions, the impulse response ϕ is a 1×1 matrix or scalar function, and via (16.14) and (16.15), there obtains

(16.16) $\phi(t) = \displaystyle\sum_{m=1}^{n} s_{im} \ s_{mj}^{(-1)} \ \exp(\lambda_m t),$

the sum-of-exponentials model for ϕ.

Let us go a step further and assume that the single input compartment is the same as the single output compartment. (For the following result we do not need to assume the eigenvalues are distinct.)

THEOREM 16.1.   [204], [119]. If the input and output compartments are the same, and if the $n \times n$ matrix A is symmetrizable by a positive definite diagonal matrix, then the impulse response function is a scalar of the form

$$\phi(t) = \sum_{k=1}^{n} \gamma_k \exp(\lambda_k t)$$

where each λ_k is real and nonpositive, and every $\gamma_k \geq 0$.

Proof.   Let the compartments be numbered so that both observation and input occur in compartment one, and only in that compartment.   Let v_i, $i = 1, 2, \ldots, n$, be eigenvectors of A corresponding to the eigenvalues λ_i.   Since A is symmetrizable by a diagonal matrix Q, the λ_i are real and so

$$\lambda_1 \leq \lambda_2 \leq \cdots \leq \lambda_n \leq 0.$$

Also due to the symmetrizability of A, we may assume that the v_i are orthonormal with respect to the Q-inner product:

$$\langle v_i, v_j \rangle_Q \equiv \langle v_i, Q v_j \rangle = \delta_{ij} .$$

The vector B,

$$B^T \equiv C \equiv [1 \quad 0 \quad \ldots \quad 0],$$

can be expanded in terms of the $\{v_i\}$,

$$B = \sum_{i=1}^{n} b_i v_i,$$

where each

$$b_i = \langle v_i, B \rangle_Q = v_{1i} q_{11}$$

in which v_{1i} is the first component of the vector v_i and $q_{11} > 0$ is the (1, 1)-entry in Q.   Now the impulse response is the 1×1 matrix

$$C \exp(tA) B = B^T \exp(tA) \sum_{i=1}^{n} b_i v_i$$

$$= \sum_{i=1}^{n} b_i B^T \exp(tA) \ v_i = \sum_{i=1}^{n} b_i B^T \exp(\lambda_i t) \ v_i$$

$$= \sum_{i=1}^{n} q_{11} \ v_{1i}^2 \ \exp(\lambda_i t),$$

which completes the proof (observe that the coefficients $q_{11} \ v_{1i}^2$ are strictly positive if and only if $v_{1i} \neq 0$).

$\boxed{///}$

16F. Nonlinear Identification Equations

Let us now address the question of just what formulas are available relating the measured impulse response parameters in ϕ (or $\hat{\phi}$) to the unknown system parameters of (A, B, C).

First, we note that information is available from the <u>characteristic polynomial of A</u>. On the one hand,

$$(16.17) \qquad \det(\lambda I - A) \equiv \lambda^n + s_1 \lambda^{n-1} + s_2 \lambda^{n-2} + \ldots + s_n,$$

where s_m is $(-1)^m$ times the sum of all the m-square principal minors of A, m = 1, 2, ..., n [17, p. 151]. On the other hand, the characteristic polynomial of A is also

$$(16.18) \qquad \prod_{i=1}^{n} (\lambda - \lambda_i) \equiv \lambda^n - \sigma_1 \lambda^{n-1} + \sigma_2 \lambda^{n-2} - \ldots + (-1)^n \sigma_n$$

where the σ s are the <u>elementary symmetric functions</u> in λ_1, λ_2, ..., λ_n [125, p. 200] and are given by

$$\sigma_1 \equiv \sum_i \lambda_i$$

$$\sigma_2 \equiv \sum_{i<j} \lambda_i \lambda_j$$

$$\sigma_3 \equiv \sum_{i<j<k} \lambda_i \lambda_j \lambda_k$$

$$\vdots \qquad \vdots$$

$$\sigma_n \equiv \lambda_1 \lambda_2 \cdots \lambda_n.$$

If corresponding coefficients are equated between (16.17) and (16.18), there results

(16.19)

$$s_1 = -\sigma_1 \qquad s_2 = \sigma_2 \qquad s_3 = -\sigma_3 \qquad \cdots \qquad s_n = (-1)^n \sigma_n$$

which are nonlinear equations in the unknown a_{ij} in terms of the measured eigenvalues λ_i of A.

Second, the $p \times q$ matrices defined by

$$M_k \equiv \phi^{(k)}(0) \qquad\qquad k = 0, 1, 2, 3, \ldots ,$$

the derivatives of the known impulse response function, are called the Markov parameters of the system (A, B, C).   Note that if ϕ has the scalar form ($p = q = 1$)

$$\phi(t) = \sum_i \gamma_i \exp(\lambda_i t)$$

then the Markov parameters are

$$\sum_i \gamma_i \lambda_i^k \qquad\qquad k = 0, 1, 2, \ldots .$$

Exercise 16.4.   If $\phi(t)$ is the impulse response of the system (A, B, C), show that

(16.20)

$$M_k = C A^k B \qquad\qquad k = 0, 1, 2, \ldots .$$

At first glance it would seem, from equation (16.20), that we have a great number of equations for the $\{a_{ij}\}$ with which to work. However, the matrices $\{M_k\}$ are not all independent, for a recursion relation can be set up as follows. Let

$$\sum_{i=0}^{n} d_i \lambda^i$$

denote the characteristic polynomial of A. Then for any $\nu = 0, 1, 2, \ldots,$

$$d_n M_{n+\nu} + d_{n-1} M_{n-1+\nu} + \ldots + d_0 M_\nu$$

$$(16.21) \qquad = d_n C A^{n+\nu} B + d_{n-1} C A^{n-1+\nu} B + \ldots + d_0 C A^\nu B$$

$$= C A^\nu [d_n A^n + d_{n-1} A^{n-1} + \ldots + d_0 I] B.$$

The expression inside the braces in (16.21) is zero by the <u>Cayley-Hamilton Theorem</u>, and therefore a recursion relation for $\{M_k\}$ obtains from (16.21).

<u>Exercise 16.5.</u>   Let e_p denote the pth column (vector) of the $n \times n$ identity matrix. For the system (A, B, C), select $B \equiv e_j$ and $C \equiv e_i^T$.   Show that

$$(16.22) \qquad M_k = a_{ij}^{(k)} \qquad\qquad k = 1, 2, \ldots$$

where $a_{ij}^{(k)}$ is the (i, j)-element of A^k.

Third, there are some <u>special formulas</u> for elements of A that can be utilized. Equation (16.22) when $k = 1$ is such a formula.   Another is expression (16.23) below. Suppose A is a SEC matrix for a compartmental system with no traps, the exit compartment is numbered m, single input is to the ith compartment so that $B = e_i$, the single input u(t) occurs over the interval [0, T] for $T \geq 0$, and the mth compartment is sampled over time (thus $C = e_m^T$).   Then the only nonzero excretion rate constant of the system is given by

$$(16.23) \qquad \boxed{ a_{0m} = \int_0^T u(\tau) \, d\tau \; / \int_0^\infty x_m(\tau) \, d\tau . }$$

This expression is called <u>Hamilton's formula</u> in the 1-dimensional case [194].   To verify it, we have the system output of

$$y(t) = Cx(t) = x_m(t),$$

which is known over time.   The sum of the columns of $\dot{x} = Ax + Bu$ is found to be

$$d(x_1 + x_2 + \ldots + x_n)/dt = - a_{0m} x_m + u$$

which in turn leads to

(16.24) $x_1(t) + \ldots + x_n(t) = - a_{0m} \int_0^t x_m(\tau)d\tau + \int_0^t u(\tau)d\tau + \gamma.$

The constant of integration γ is zero since $x(0) = 0$. Because the compartmental system is <u>open</u>, has no traps, and $u(t) \equiv 0$ for $t > T$, then each $x_j(t)$ must approach zero as $t \to \infty$. Hence, in the limit, we see from (16.24) that

$$0 = - a_{0m} \int_0^\infty x_m(\tau)d\tau + \int_0^T u(\tau)d\tau$$

from which formula (16.23) obtains.

<u>Exercise 16.6</u>.   Give an alternate derivation of (16.23) using the fact that each entry in the mth row of A^{-1} is $- 1/a_{0m}$.

One advantage of formula (16.23) is that the integral of the denominator can be computed <u>numerically</u> directly from data, thus circumventing the curve-fitting problem for x_m. We had a similar situation in equation (14.9) for the mean time τ_{im}. However, if this approach is taken, the new problem of truncation error arises. Since data are gathered only over a finite time interval $[0, t_d]$, the integral must be approximated by

$$\int_0^{t_d} x_m(\tau)d\tau .$$

Usually, however, t_d is large enough so that $x(t_d) \approx 0$ and hence the error

$$\int_{t_d}^\infty x_m(\tau)d\tau$$

is quite small.

<u>Exercise 16.6</u>.   [37], [245], [110]. A tracer study is done on a physiological system modeled as in Figure 16.2. Compartment one receives a bolus injection of a unit amount of tracer at $t = 0$, and that same compartment is sampled over a period of ten days with the results presented in Table 16.1. The main system parameter

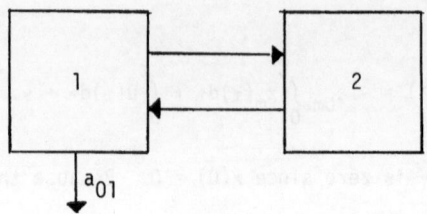

Figure 16.2.

A Two-Compartment SEC System

of clinical interest is a_{01}. Compute this excretion rate constant by Hamilton's
formula (16.23).

TABLE 16.1.

t	0.5	1.	1.5	2.	3.	4.
x_1	0.461	0.259	0.170	0.121	0.0722	0.0451

t	5.	6.	7.	8.	9.	10.
x_1	0.0319	0.0240	0.0182	0.0141	0.0100	0.0094

Formula (16.23) has many other applications, one of which is in the "indicator-
dilution method" of measurement of the volume of human blood or the mean blood flow
rate through the cardiovascular system [194, p. 111]. Also see the model of
Berkstresser et al. [29] for deriving a measure of the renal function.

16G. A Three Compartment Model

Let us illustrate the use of formulae in (16.19), (16.20), and (16.23) on the
three compartment model depicted in Figure 16.3. The associated compartmental
matrix is

$$A = \begin{bmatrix} a_{11} & 0 & a_{13} \\ a_{21} & a_{22} & 0 \\ 0 & a_{32} & a_{33} \end{bmatrix} \qquad \begin{aligned} a_{11} &\equiv -a_{21} \\ a_{22} &\equiv -a_{32} \\ a_{33} &\equiv -a_{13} - a_{03} \end{aligned} .$$

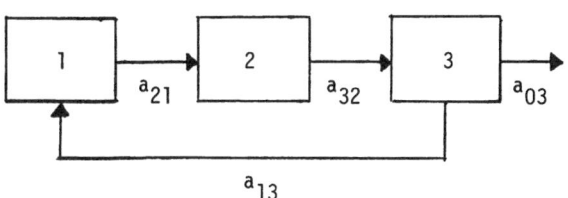

Figure 16.3.

Suppose that at time zero, a known bolus amount of D units of tracer is injected into compartment two and subsequently the third compartment is sampled over time. We take

$$B \equiv [0 \quad 1 \quad 0]^T \qquad\qquad C \equiv [0 \quad 0 \quad 1]$$

with single input function $u(t) = D\delta(t)$ and observed output function $y(t) = Cx(t) = x_3(t)$. From equation (16.10), we have the (assumed known) impulse response function

$$(16.24) \qquad \phi(t) = y(t)/D = x_3(t)/D.$$

Since the output function $x_3(t)$ is a component of the solution of a system of linear differential equations, we would expect to curve-fit the form

$$x_3(t) \equiv \sum_{i=1}^{3} A_i \exp(\lambda_i t)$$

to the discrete data observations for x_3 (see Section 23). Thus (16.24) implies that

$$(16.25) \qquad \phi(t) = \sum_{i=1}^{3} \gamma_i \exp(\lambda_i t), \qquad \gamma_i = A_i/D.$$

Assuming distinct real eigenvalues λ_i, which the data should indicate, the form of ϕ in (16.25) is in agreement with (16.16).

Let us assume that a nonlinear least squares curve-fitting of x_3 has been performed over the time interval of measurement so that the parameters A_i and λ_i are now known. Hence from (16.25) the impulse response parameters γ_i and λ_i are known. We proceed to (uniquely) identify the entries of A if possible. Thus the question is posed: Is knowledge of the estimates γ_1, γ_2, γ_3, λ_1, λ_2, and λ_3 in ϕ sufficient to yield unique determination of all unknown a_{ij} in A?

Of the four unknown system parameters a_{21}, a_{32}, a_{13}, and a_{03}, the last can be computed through Hamilton's formula (16.23):

$$a_{03} = \int_0^T u(\tau)d\tau / \int_0^\infty x_3(\tau)d\tau = - D/(A_1/\lambda_1 + A_2/\lambda_2 + A_3/\lambda_3).$$

Here we have used the delta function property of [111]

$$\int_0^T u(\tau)d\tau = \int_{-\infty}^\infty D\delta(\tau)d\tau = D.$$

Hence the single excretion rate constant, in terms of the impulse response parameters, is given by

$$a_{03} = - [\gamma_1/\lambda_1 + \gamma_2/\lambda_2 + \gamma_3/\lambda_3]^{-1} .$$

The first Markov parameter $M_1 = CAB = a_{32}$ as treated in (16.20) and (16.22) yields a formula for the fractional transfer coefficient

$$a_{32} = \dot{\phi}(0) = \sum_{i=1}^3 \gamma_i \lambda_i .$$

It remains to determine a_{21} and a_{13}. These can be found from the characteristic polynomial of the matrix A. The last equation of (16.19) is

$$- a_{21} a_{32} a_{03} = \det A = \lambda_1 \lambda_2 \lambda_3 .$$

Since a_{03} and a_{32} have already been determined, the previous equation allows the calculation

$$a_{21} = - \lambda_1 \lambda_2 \lambda_3 / a_{32} a_{03} .$$

Finally, a_{13} is found through the first equation in, (16.19), which says that the trace of a matrix is equal to the sum of the eigenvalues of that matrix:

$$a_{11} + a_{22} + a_{33} = \lambda_1 + \lambda_2 + \lambda_3$$

or

$$a_{13} = -\lambda_1 - \lambda_2 - \lambda_3 - a_{21} - a_{32} - a_{03} .$$

Thus we see that the three compartment model of Figure 16.3 is identifiable, for the above equations do in fact furnish unique solutions for a_{03}, a_{32}, a_{21}, and a_{13} in terms of the curve-fitted impulse response parameters γ_i and λ_i.

16H. A Four Compartment Model

Let us consider another example of the identification process. References [9], [199] treat the mathematical model for tracer transport dealing with clinical differentiation between major categories of liver disease in infants (also see Section 26). To aid in this diagnostic process, a test based on the disappearance rate of the radioactive tracer Rose Bengal is devised. The physiological basis for the model centers on the transfer of bilirubin. It is the excess of bilirubin in the blood and body tissues that causes jaundice, the yellow discoloration associated with hepatic disease. The idealized pathway of tracer transport through the biliary system is illustrated as the closed system of Figure 16.4 in which compartment one is injected with a known amount D of tracer at time zero, and the first, second, and fourth compartments are measured in time. Thus the equations for the linearized tracer experiment are

$$\dot{x} = Ax + Bu$$

$$x(0) = 0$$

$$y = Cx$$

where

$$A = \begin{bmatrix} a_{11} & 0 & a_{13} & 0 \\ a_{21} & a_{22} & 0 & 0 \\ a_{31} & 0 & a_{33} & 0 \\ 0 & 0 & a_{43} & a_{44} \end{bmatrix} \qquad \begin{array}{l} a_{11} = -a_{21} - a_{31} \\ a_{22} = 0 \\ a_{33} = -a_{13} - a_{43} \\ a_{44} = 0 \end{array} .$$

$$B = [1 \quad 0 \quad 0 \quad 0]^T \qquad C = \begin{bmatrix} 1 & 0 & 0 & 0 \\ 0 & 1 & 0 & 0 \\ 0 & 0 & 0 & 1 \end{bmatrix} .$$

$$u(t) = D\delta(t)$$

Under this experimental setup, we ask whether or not the system matrix A is identifiable.

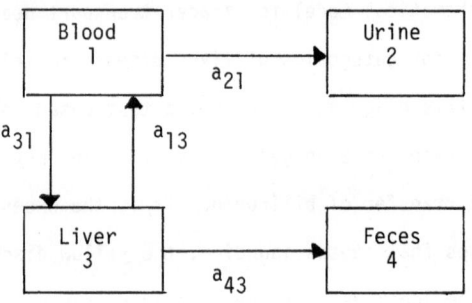

Figure 16.4.

Pathway of Rose Bengal

Since the single input of Rose Bengal into the system is $u(t) = D\delta(t)$, the sampled output function is

$$D\phi = y = Cx = [x_1 \quad x_2 \quad x_4]^T$$

from equation (16.10). Hence the 3×1 vector

$$\phi(t) = [x_1(t)/D \qquad x_2(t)/D \qquad x_4(t)/D]^T$$

is assumed to have known entries for $t \geq 0$. As it turns out, the a_{ij}s can be identified through the Markov parameters $M_k \equiv \phi^{(k)}(0)$. For, from (16.20),

$$\phi^{(1)}(0) = M_1 = CAB = [a_{11} \quad a_{21} \quad 0]^T,$$

and thus a_{11} and a_{21} are known. Since $a_{11} = - a_{21} - a_{31}$, then a_{21} and a_{31} are identified:

$$a_{21} = \dot{x}_2(0)/D$$

$$a_{31} = -a_{21} - a_{11} = -a_{21} - \dot{x}_1(0)/D.$$

The remaining two parameters are found via the second Markov parameter M_2. As before, let $a_{ij}^{(k)}$ be the (i, j)-element of A^k. Then

$$a_{11}^{(2)} = a_{11}^2 + a_{13}a_{31}$$

$$a_{21}^{(2)} = a_{21}a_{11}$$

$$a_{41}^{(2)} = a_{43}a_{31}$$

and so

$$\phi^{(2)}(0) = M_2 = C A^2 B = [a_{11}^{(2)} \quad a_{21}^{(2)} \quad a_{41}^{(2)}]^T$$

(16.26)

$$= [a_{11}^2 + a_{13}a_{31} \quad a_{21}a_{11} \quad a_{43}a_{31}]^T .$$

The first component of this vector yields the identity of a_{13},

$$a_{13} = (\ddot{x}_1(0)/D - a_{11}^2)/a_{31} ,$$

since a_{11} and a_{31} are already known. Moreover, equality of the third components in (16.26) gives the formula

$$a_{43} = \ddot{x}_4(0)/Da_{31}$$

completing the identification of A.

Exercise 16.7. Discuss the identification of the A matrix for the tracer system (A, B, C) where

$$A = \begin{bmatrix} a_{11} & a_{12} \\ a_{21} & a_{22} \end{bmatrix} \qquad \begin{array}{l} a_{11} = -a_{21} \\ \\ a_{22} = -a_{12} - a_{02} \end{array}$$

$$.B = [1 \quad 0]^T \qquad\qquad C = [1 \quad 1]$$

with the bolus input hypothesis.

Exercise 16.8. Consider a three compartment SEC mammillary system with the mother compartment also being the single exit compartment of the system. State experimental conditions under which the system matrix is identifiable.

Exercise 16.9. A major use of compartmental modeling is in pharmacokinetics. The origins of this type of analysis go back to Teorell's work in 1937 in which he analyzed the three compartment model shown in Figure 16.5 [47]. With bolus input at time zero into compartment one, and with subsequent time measurement of compartments one and two, discuss the identification of the system matrix $A \equiv [a_{ij}]$.

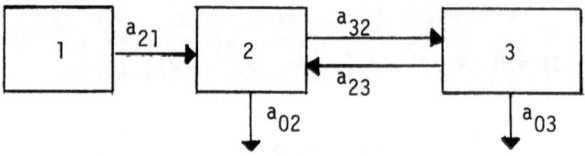

Figure 16.5

Pharmacokinetic Model of Teorell

Exercise 16.10. Consider the four compartment SEC model pictured in Figure 16.6 in which compartment 1 has a bolus injection of tracer at time zero, and is then sampled over time. Is the system matrix identifiable? (This model has been used for the analysis of calcium and strontium metabolism in man [190]).

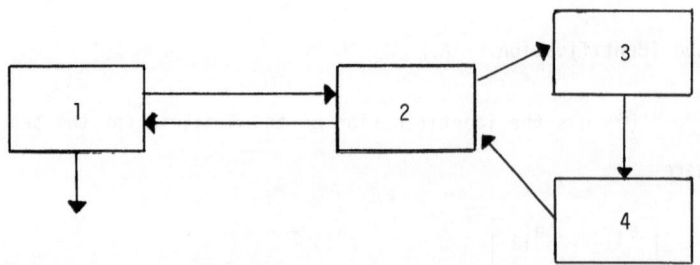

Figure 16.6

Upon reviewing the above examples and exercises, two chief difficulties in identification of $A \equiv [a_{ij}]$ in (16.5) are seen [82]:

(a) the appropriate nonlinear equations in the $\{a_{ij}\}$ can be hard to find;

(b) when these equations are known, it must be determined whether they have a unique positive solution $\{a_{ij}^*\}$, and if so, that solution must be computed.

SECTION 17. NECESSARY AND SUFFICIENT CONDITIONS FOR IDENTIFIABILITY

17A. Model Identifiability

Let us examine the problem of structural identification more closely. We are saying that if, in an identification experiment, the following are known:

(a) the system output function y;

(b) the number of compartments;

(c) the connectivity of the system (i.e., which fractional transfer coefficients are nonzero);

(c) the initial conditions;

(d) the inputs;

then the model is (uniquely) <u>identifiable</u> if all of its fractional transfer coefficients a_{ij} (i $\neq$ j) can be uniquely determined as positive values from ideal or error free data. Perfect data is a prerequisite for consistent parameter estimates when using actual experimental data. Since all of the system parameters appear in the impulse response function ϕ (or in $\hat{\phi}$), the problem of identifying A from experimental measurements of the system output y is reduced to identifying A from ϕ or $\hat{\phi}$ (either of which is considered as known). Given the system (A, B, C), the impulse response function ϕ (or the transfer function $\hat{\phi}$) is precisely determined. However, given ϕ, it is generally possible to find <u>many</u> systems (A, B, C) that generate the function ϕ, i.e., that are <u>realizations of ϕ</u> (see Section 21).

17B. Necessary Conditions

Necessary conditions for identifiability can be developed in terms of the connectivity diagram for the compartmental system [64], [75], [214], [8], [135]. Tracer inputs are injected into certain compartments of the system and then the output time course y(t) of some compartments is observed. To have a possibility of determining a particular a_{ij}, the transfer of material from compartment j to compartment i <u>must affect</u> the measured compartments. Consider the compartmental system whose connectivity diagram is shown in Figure 17.1. Suppose compartment 1 is injected with tracer and compartment 5 is sampled. All transfers of tracer to or from compartments which are on any path from 1 to 5 can influence the output y.

Thus a_{21} affects compartment 1 but a_{32} does not. Since no tracer can pass to compartment 6, none can then pass to compartment 5 and so a_{56} does not influence y.

Because a_{32} and a_{56} cannot affect the observation y, they are <u>nonidentifiable</u> and hence the system is nonidentifiable (under the present experiment).

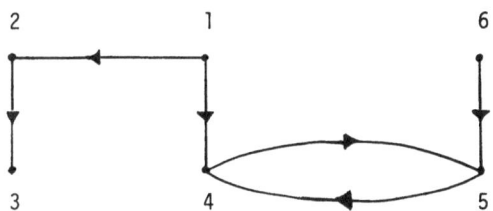

Figure 17.1.

Compartment k is input reachable if there is at least one input component that reaches k. If each compartment in the system is input reachable, then the system is said to be <u>input reachable</u>. All compartments in Figure 17.1 are input reachable from 1 except 6; thus the system is not input reachable. A little thought will bear out the fact that if a compartment k is not input reachable then any model parameter a_{ik} are a_{kj} is not going to be identifiable.

Compartment k is output reachable if there is a path from k to an observed compartment. The system is said to be <u>output reachable</u> provided every compartment in the system is output reachable. The system in Figure 17.1 is not output reachable because there is no path from compartment 3 to sampled compartment 5. These conditions can be checked by examining the reachability matrix of the connectivity diagram of the system [214], [136].

In contrast to input reachability, it is possible to structurally identify certain compartmental systems even if the system is not output reachable. For example, consider the system depicted in Figure 17.2 in which compartment 2 is not output reachable if compartment 1 is bolus input and also subsequently sampled. The system matrices are

$$A = \begin{bmatrix} a_{11} & 0 \\ a_{21} & a_{22} \end{bmatrix} \qquad \begin{matrix} a_{11} = -a_{21} \\ \\ a_{22} = -a_{02} \end{matrix} \qquad C = [1 \quad 0] = B^T$$

and so the transfer function is

$$(17.1) \qquad \hat{\phi}(s) = C(sI-A)^{-1} B = (s-a_{22})/(s^2 - (a_{11} + a_{22})s + a_{11}a_{22}) .$$

Because this last expression simplifies to $1/(s-a_{11})$, only a_{11} is identifiable and not both a_{11} and a_{22} as might have been originally thought from the form of $\hat{\phi}$. Thus if a_{02} is nonzero, the system is nonidentifiable. However, if the additional constraint $a_{02} = 0$ is imposed, then the system is fully identifiable.

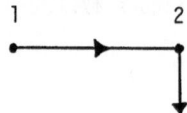

Figure 17.2.

The above definitions and examples suggest the following result, which is readily tested from the connectivity diagram of the system [135], [136].

THEOREM 17.1. In order that an input-output tracer experiment identify all the fractional transfer coefficients it is necessary that (a) all compartments are input reachable; (b) all compartments possessing at least one path leaving them (including excretions) must be output reachable.

Stronger necessary conditions developed by Cobelli et al. [64], and refined by Anderson [6], are the primary subject of Section 18.

17C. Sufficient Conditions

Supposing the necessary conditions are fulfilled, some of the fractional transfer coefficients in the system still may not be identifiable. The impulse response function ϕ or the transfer function $\hat{\phi}$ must be examined to see whether all of the a_{ij} appear in these functions and if so, whether all are uniquely determined.

The number μ of unknown parameters a_{ij} must be compared with the number ν of inde-pendent coefficients involving the a_{ij} available from the known ϕ or $\hat{\phi}$.

Let us consider the equations available from knowledge of the coefficients of the transfer function (in contrast to the equations of Section 16)

$$\hat{\phi}(s) \equiv C(sI-A)^{-1}B.$$

Recall that the inverse matrix

$$(sI-A)^{-1} = adj(sI-A)/det(sI-A)$$

in which adj refers to the adjoint matrix of sI-A, and det(sI-A) is the character-istic equation (16.17) of A, where the coefficients s_m (m = 1,2,...,n) of (16.17) are functions of the fractional transfer coefficients a_{ij} in A. The adjoint matrix can also be expanded as [57]

$$adj(sI-A) = R_0 s^{n-1} + R_1 s^{n-2} + \ldots + R_{n-1}$$

in which each R_i is an $n \times n$ matrix with entries that are functions of the a_{ij}, in particular

$$R_0 = I$$

$$R_1 = A + s_1 I$$

$$R_2 = A^2 + s_1 A + s_2 I$$

$$\vdots \qquad \vdots$$

$$R_{n-1} = A^{n-1} + s_1 A^{n-2} + \ldots + s_{n-1} I.$$

Thus the transfer function has the representation

$$\hat{\phi}(s) = C[R_0 s^{n-1} + R_1 s^{n-2} + \ldots + R_{n-1}]B/det(sI-A)$$

$$= \{[CB]s^{n-1} + [CAB + s_1 CB]s^{n-2} + [CA^2 B + s_1 CAB + s_2 CB]s^{n-3} + \ldots$$

$$+ [CA^{n-1}B + s_1 CA^{n-2}B + \ldots + s_{n-1}CB]\}/det(sI-A)$$

which can be expressed in terms of the Markov parameters as

$$\hat{\phi}(s) = \{M_0 s^{n-1} + [M_1 + s_1 M_0] s^{n-2} + [M_2 + s_1 M_1 + s_2 M_0] s^{n-3} + \dots$$

(17.2)

$$+ [M_{n-1} + s_1 M_{n-2} + \dots + s_{n-1} M_0]\}/\det(sI-A).$$

Hence the transfer function is a rational function in s, where the numerator is a polynomial of degree at most n-1, and where the denominator is a polynomial of degree n.

Exercise 17.1. Use formula (17.2) to compute the transfer function for the system in Exercise 16.1.

Exercise 17.2. If (A, B, C) is a realization of the rational matrix function in (17.2), show that

(17.3) $\hat{\phi}(s) = M_0 s^{-1} + M_1 s^{-2} + M_2 s^{-3} + \dots$

where $M_\ell = CA^\ell B$, $\ell = 0, 1, 2, \dots$. For what values of s does the series converge?

The problem of identifying A now becomes one of examining all of the coefficients of the terms in both the numerator and denominator of $\hat{\phi}$ to check if all a_{ij} ($i \neq j$) are present and uniquely determined as positive quantities from knowledge of the coefficients. This generally is difficult to do for a system with a moderate to large number of compartments since this is a problem in simultaneous nonlinear equations.

Assuming B and C are known, so that M_0 is known, then (17.2) yields the numerator equations

$$M_1 + s_1 M_0 \equiv \Gamma_1$$

(17.4) $$M_2 + s_1 M_1 + s_2 M_0 \equiv \Gamma_2$$

$$\vdots \qquad \qquad \vdots$$

$$M_{n-1} + s_1 M_{n-2} + \dots + s_{n-1} M_0 \equiv \Gamma_{n-1}$$

along with the denominator equations (16.19)

$$s_1 = -\sigma_1$$

(17.5)
$$s_2 = \sigma_2$$

$$\begin{array}{cc} . & . \\ . & . \\ . & . \end{array}$$

$$s_n = (-1)^n \sigma_n$$

in which the Γ_i are known $p \times q$ matrices, and the σ_i are known scalars. Thus the following sufficient condition for identifiability of A $\underline{\text{in terms of}}$ $\hat{\phi}$ obtains.

THEOREM 17.2.   The $n \times n$ compartmental system matrix A is structurally identifiable if the system of nonlinear equations (17.4) and (17.5) has a unique positive solution for all nonzero a_{ij} $(i \ne j)$ in A.

Another sufficient condition for identifiability of A, but this time $\underline{\text{in}}$ $\underline{\text{terms of}}$ ϕ, can be obtained from the Markov parameters and recursive formula (16.21).

THEOREM 17.3.   The matrix A is structurally identifiable provided the set of non-linear equations

$$\phi^{(1)}(0) = CAB$$

$$\phi^{(2)}(0) = CA^2B$$

$$\begin{array}{cc} . & . \\ . & . \\ . & . \end{array}$$

$$\phi^{(n-1)}(0) = CA^{n-1}B$$

together with equations (17.5) have a unique positive solution for all nonzero a_{ij} $(i \ne j)$ in A.

Exercise 17.3.   In view of Theorems 17.2 and 17.3, reconsider any conclusions drawn in Exercises 16.7 - 16.10.

Alternate sufficient conditions for identifiability using component and modal matrices are presented in Section 20. In Section 21, a third set of sufficient conditions, which involves similarity transformations and equivalent models, is formulated.

SECTION 18.   A SIMPLE TEST FOR NONIDENTIFIABILITY

18A.   Counting Nonzero Transfer Function Coefficients

It would be very useful to have a simple test that could be easily applied to a given tracer experiment to decide in advance whether or not the given experiment is feasible based on its possible identifiability.   Such a test can be developed through the transfer function (17.2).

The number μ of parameters a_{ij} estimable is governed by the number ν of independent nonzero coefficients in the numerator and denominator of (17.2).   At least two situations can occur: (1) dependency among impulse response parameters M_i and s_j can be present in a particular model in such a way that certain coefficients in (17.2) will turn out to be zero; (2) common factors between the numerator and denominator of (17.2) can cancel, as illustrated by the model whose transfer function is (17.1), with a resulting loss in expressions involving the a_{ij}.

Through (17.2), upper bounds on ν can be constructed, along lines indicated by Chen [57, p. 301], Cobelli et al. [64], [63], [61], and Anderson [6].   The denominator of $\hat{\phi}$ contains n scalar coefficients s_1, s_2, ..., s_n involving the a_{ij}.   The numerator has n-1 terms, as in (17.4), which deal with the unknown a_{ij}.   But each of these numerator expressions is a $p \times q$ matrix.   Hence the maximal number of equations which involve the a_{ij} present as coefficients in $\hat{\phi}$ is

(18.1)   (n-1)pq + n.

This maximal value can be trimmed by observing that for a given input-output experiment and for a prescribed configuration of compartments, certain coefficients in $\hat{\phi}$ will vanish. We start by considering the case when m distinct subsystems of the compartmental system are traps.   Then the last m eigenvalues in the list

$$|\lambda_1| \geq |\lambda_2| \geq \cdots \geq |\lambda_n|$$

are zero [97].   But then the last m coefficients in the characteristic equation of A are zero by equations (16.18) and (16.19).   For instance, if n = 4 and m = 2 as in the liver disease model of Figure 16.4, then from the elementary symmetric functions

(16.19) there results

$$s_1 = - (\lambda_1 + \lambda_2 + \lambda_3 + \lambda_4)$$

$$s_2 = \lambda_1\lambda_2 + \lambda_1\lambda_3 + \lambda_1\lambda_4 + \lambda_2\lambda_3 + \lambda_2\lambda_4 + \lambda_3\lambda_4$$

$$s_3 = - (\lambda_1\lambda_2\lambda_3 + \lambda_1\lambda_2\lambda_4 + \lambda_1\lambda_3\lambda_4 + \lambda_2\lambda_3\lambda_4)$$

$$s_4 = \lambda_1\lambda_2\lambda_3\lambda_4$$

and so clearly $s_3 = s_4 = 0$ since $\lambda_3 = \lambda_4 = 0$. Moreover, Hearon [123] has shown that these remaining s_i are positive (also see Theorem 22.1). Hence the last term, n, of (18.1) can be replaced by the quantity $n - m$.

The first term in (18.1) comes from the numerator of $\hat{\phi}$. The list of coefficients there involve M_0, M_1, ..., M_{n-1}. Let us now list all input compartments as I_1, I_2, ..., I_q, and all output or sampled compartments as 0_1, 0_2, ..., 0_p. Then for each possible pair

$$\{(0_i, I_j)\} \qquad i = 1, 2, \ldots, p; \qquad j = 1, 2, \ldots, q,$$

there is a corresponding <u>shortest path length</u> $p(0_i, I_j)$ in the connectivity diagram of the system from I_j to 0_i; that is, $p(0_i, I_j)$ is the minimum number of edges travelled in the diagram in passing to node 0_i from node I_j. If $0_i = I_j$, then $p(0_i, I_j) \equiv 0$. If I_j cannot reach 0_i, then $p(0_i, I_j)$ is not defined.

<u>Exercise 18.1</u>. Prove that the (i, j)-position in M_ℓ will be zero for all $\ell = 0, 1, 2, \ldots, p(0_i, I_j) - 1$. If I_j cannot reach 0_i, then that position will be zero for all ℓ.

<u>Example 1</u>. Consider the diagram of Teorell (Figure 16.5) in which we take $a_{03} = 0$ and

$$A = \begin{bmatrix} a_{11} & 0 & 0 \\ a_{21} & a_{22} & a_{23} \\ 0 & a_{32} & a_{33} \end{bmatrix} \qquad \begin{array}{l} a_{11} = - a_{21} \\[1ex] a_{22} = - a_{32} - a_{02} \\[1ex] a_{33} = - a_{23} \end{array}$$

$$B = [1 \quad 0 \quad 0]^T \qquad C = \begin{bmatrix} 1 & 0 & 0 \\ 0 & 0 & 1 \end{bmatrix}.$$

The transfer function $\hat{\phi}$ is computed via formula (17.2). Directly from A and its principal minors, we get

$$s_1 = - (a_{11} + a_{22} + a_{33})$$

(18.2) $\qquad s_2 = a_{11}a_{22} + a_{11}a_{33} + a_{02}a_{23}$

$$s_3 = - a_{11}a_{02}a_{23} .$$

As there are no traps in the system, $m = 0$ and so s_1, s_2, $s_3 > 0$. Denote the input and output compartments as

$$I_1 = 1, \quad O_1 = 1, \quad O_2 = 3 .$$

From the connectivity diagram we see that

$$p(O_1, I_1) = 0, \quad p(O_2, I_1) = 2.$$

Thus the (2, 1)-entry in M_0 and M_1 should be zero according to Exercise 18.1. The Markov parameters are the 2×1 matrices

$$M_0 = CB \quad = [1 \qquad 0]^T$$

$$M_1 = CAB = [a_{11} \qquad 0]^T$$

$$M_2 = CA^2B = [a_{11}^{(2)} \quad a_{31}^{(2)}]^T .$$

Hence the numerator coefficients in $\hat{\phi}$ in order of decreasing powers of s are

$$M_0 \quad = [1 \qquad 0]^T$$

(18.3) $\qquad M_1 + s_1 M_0 \; = \; [a_{11} + s_1 \qquad 0]^T$

$$M_2 + s_1 M_1 + s_2 M_0 = [a_{11}^{(2)} + s_1 a_{11} + s_2 \qquad a_{31}^{(2)}]^T .$$

Reading the numerator in the (1, 1)-entry of $\hat{\phi}$ we see at most $3 - p(0_1, I_1) = 3$ nonzero coefficients, and from scanning the numerator of the (2, 1)-element of $\hat{\phi}$, there is at most $3 - p(0_2, I_1) = 1$ nonzero coefficient.

By building the above considerations into our general expression, formula (18.1) is further refined. Now if $0_i = I_j$ for some i and j, then $p(0_i, I_j) = 0$ and the coefficient of s^{n-1} in the numerator of the (i, j)-entry in $\hat{\phi}$ will be unity. These coefficients of s^{n-1} come from $M_0 = CB$, which does not contain any a_{ij} and hence should not be counted as an expression containing the a_{ij}. Thus we adopt the following rule: If $0_i = I_j$, so that $p(0_i, I_j) = 0$, then the corresponding term [·] in expression (18.4) is reduced by one. Now the maximal number of nonzero coefficients in $\hat{\phi}$ which involve the unknown a_{ij} of A is given by

$$(18.4) \qquad \sum_{i=1}^{p} \sum_{j=1}^{q} [n - p(0_i, I_j)] + (n - m) .$$

If 0_i is not reachable from I_j, then the corresponding value in the brackets [·] for that term of (18.4) is defined to be zero.

Exercise 18.2.   Show that Teorell's model with $a_{02} = 0$, injection to compartments 1 and 3, and sampling of only compartment 1 illustrates the last comment. The matrix $\hat{\phi}$ is 1×2 but its second entry is zero. If we choose $0_1 = 1$, $I_1 = 1$, and $I_2 = 3$, then I_2 cannot reach 0_1 and this is reflected in (18.4) by the term $n - p(0_1, I_2)$ being equal to zero. That is, the (1, 2)-entry of $\hat{\phi}$ adds no new equations for the a_{ij} in the system (17.4) and (17.5).

Example 1 (continued).   To demonstrate formula (18.4), we return to (18.2) and (18.3). By actual count, there are six nonzero expressions in $\hat{\phi}$ which involve the a_{ij}:

$$a_{11} + s_1 \qquad\qquad\qquad s_1$$

$$a_{11}^{(2)} + s_1 a_{11} + s_2 \qquad\qquad s_2$$

$$a_{31}^{(2)} \qquad\qquad\qquad\qquad s_3$$

By contrast, formula (18.4) - which does not require the calculation of $\hat{\phi}$ - predicts that the maximal number of nonzero coefficients in $\hat{\phi}$ is

$$\sum_{i=1}^{2} \sum_{j=1}^{1} [3 - p(0_i, I_j)] + (3 - 0) = [3 - 1] + [3 - 2] + 3 = 6.$$

18B.   Coefficient Structure

Besides the coefficients of highest order possibly being zero, the coefficients in the numerator of $\hat{\phi}$ of lowest order may also vanish.   The lowest order coefficient of (17.2) is the constant term

(18.5) $M_{n-1} + s_1 M_{n-2} + s_2 M_{n-3} + \ldots + s_{n-1} M_0$.

We can preclude this from having the value zero in certain cases [6].

THEOREM 18.1   Suppose the compartmental system has no traps.   Then the coefficient (18.5) is the nonnegative number

$$s_n \int_0^\infty \phi(t)dt .$$

If input compartment I_j can reach output compartment 0_i, then the (i, j)-entry of (18.5) is positive.   If I_j cannot reach 0_i, then the (i, j)-entry of (18.5) is zero.

Proof.   The Cayley-Hamilton theorem states that the system matrix A satisfies its own characteristic equation:

(18.6) $A^n + s_1 A^{n-1} + s_2 A^{n-2} + \ldots + s_n I = 0.$

Since the system has no traps, the compartmental matrix is invertible [97] and $s_n = (-1)^n \det A > 0$ [123].   Multiply (18.6) by A^{-1} and move the last term to the right-hand side of the equation:

(18.7) $A^{n-1} + s_1 A^{n-2} + s_2 A^{n-3} + \ldots + s_{n-1} I = - s_n A^{-1}.$

Upon premultiplying (18.7) by C and postmultiplying by B, we have

(18.8) $M_{n-1} + s_1 M_{n-2} + \ldots + s_{n-1} M_0 = s_n C(-A^{-1})B.$

But every factor on the right-hand side of (18.8) is positive or nonnegative (see Section 13). Hence the left-hand side of (18.8) is nonnegative. Because A is invertible and the real part of any of its eigenvalues is negative, then

$$(18.9) \qquad \int_0^\infty \phi(t)dt = C \int_0^\infty \exp(tA)dt \, B = C(-A^{-1})B$$

where the last equality comes from Exercise 14.2. Thus the left-hand side of (18.8) is equal to s_n multiplied by the integral of the impulse response function. If I_j reaches 0_i, then the $(0_i, I_j)$-entry of A^{-1} is negative by the inequalities of (13.10). Hence the (i, j)-element on the right-hand side of (18.8) is positive. If I_j cannot reach 0_i, then the mean time

$$- a_{0_i I_j}^{(-1)} = 0$$

and so the (i, j)-entry in (18.8) is zero.

▨

Exercise 18.3. Show that the n coefficients Γ_i of (17.4) in the numerator of $\hat\phi(s)$, for a compartmental system with no traps, are given (in order of decreasing powers of s) by

$$(-1)^n \; \Gamma_0 = (-1)^n s_1 CTB + (-1)^{n-1} s_2 CT^2 B + \ldots + (-1)s_n CT^n B$$

$$(-1)^{n-1}\Gamma_1 = (-1)^{n-1} s_2 CTB + (-1)^{n-2} s_3 CT^2 B + \ldots + (-1)s_n CT^{n-1}B$$

(18.10)

$$\vdots$$

$$(-1)\Gamma_{n-1} = (-1)s_n CTB$$

where $T \equiv - A^{-1}$ (see Section 14).

Exercise 18.4. A number of the coefficients in $\hat\phi$ have been observed to be non-negative. Prove or disprove: all coefficients in $\hat\phi$ are nonnegative. (See Section 22).

The combinations $CT^\ell B$, seen above in formulas (18.10), can be further characterized as moments of the impulse response function, a special case of which has already been derived in (18.9).

THEOREM 18.2.   Suppose the compartmental system has no traps. Then

(18.11) $CT^{\ell+1}B = \int_0^\infty t^\ell \phi(t)dt/\ell!$ $\ell = 0, 1, 2, \ldots$.

Proof.   Since the system has no traps, then A is invertible and any eigenvalue of A will have a strictly negative real part. Consider

(18.12) $(-1)^{\ell+1} \int_0^\infty t^\ell e^{\lambda t}dt/\ell!$ $Re(\lambda) < 0$.

Successive integration by parts of the integral shows that (18.12) is equal to $\lambda^{-\ell-1}$. Hence the corresponding function of the matrix is [143, p. 183]

$$(-1)^{\ell+1} A^{-\ell-1} = \int_0^\infty t^\ell e^{At}dt/\ell! \; .$$

Equation (18.11) then obtains immediately upon multiplying the last equation by C and B.

$\blacksquare$

Exercise 18.5.   If the compartmental system contains no traps, verify that

$$\hat{\phi}(s) = \sum_{j=1}^n CT^j B(-s)^{j-1}$$

in a sufficiently small neighborhood of s = 0.

One observation coming out of the last few theorems and exercises is that the impulse response function, the transfer function, the Markov parameters, and the $CT^\ell B$ are all mutually computable from one another.

Let us return to the coefficient (18.5) in the numerator of $\hat{\phi}$. Via Theorem 18.1 it is seen that if input compartment I_j can reach output compartment 0_i, then this coefficient in $\hat{\phi}_{ij}$ is nonzero in the case of a system with no traps. Moreover, it can be shown from (17.2) that intermediate powers between

$$s^{n-p(0_i, I_j)-1}$$

and s^0 (inclusive) will not be zero under these conditions. Therefore the maximal number of nonzero coefficients in $\hat{\phi}$ for a compartmental system with no traps is given by (18.4) with m = 0.

18C. Further Refinements of Formula (18.4)

We proceed to the case where the compartmental system has $m \geq 1$ traps. The transfer function $\hat{\phi}$ is made up of components $\hat{\phi}_{ij}$ each of which is the transfer function between input compartment I_j and output compartment O_i, and thus refers to the subsystem of the compartmental system made up of the number $n(O_i, I_j)$ of those compartments which are output reachable to O_i as well as input reachable from I_j. For instance, consider the system in Figure 18.1 in which compartment 1 is input, and compartments 1 and 2 are measured in time. Define $I_1 \equiv 1$, $O_1 \equiv 1$, and $O_2 \equiv 2$.

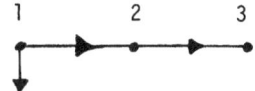

Figure 18.1.

For the pair (O_1, I_1), the intersection of the set of input reachable compartments with the set of output reachable compartments is $\{1, 2, 3\} \cap \{1\} = \{1\}$ so that $n(O_1, I_1) = 1$. For the pair (O_2, I_1) we likewise have $\{1, 2, 3\} \cap \{1, 2\} = \{1, 2\}$ from which $n(O_2, I_1) = 2$.

Note that if all compartments of a system are output reachable to O_i as well as input reachable from I_j, then $n(O_i, I_j) = n$. On the other hand, if the system has traps, as we are presently considering, then in general $n(O_i, I_j) \leq n$ as demonstrated by the model in Figure 18.1.

Our analysis with $p(O_i, I_j)$ treated the leading coefficients in the numerator of $\hat{\phi}_{ij}$ which are zero. Now we consider the trailing coefficients which may be equal to zero. For the subsystem made up of the $n(O_i, I_j)$ compartments, let $q(O_i, I_j)$ be the number of coefficients of lowest order in $\hat{\phi}_{ij}$ which equal zero; that is, the coefficients of the powers s^v in the numerator of $\hat{\phi}_{ij}$ equal to zero for

$$0 \leq v \leq q(O_i, I_j) - 1.$$

As previously mentioned, the last m coefficients of the characteristic polynomial of A are zero, and the first $n - m$ coefficients are positive. Hence the common denominator $\det(sI - A)$ in $\hat{\phi}$ carries the factor s^m. The following result tells us that the numerator of $\hat{\phi}_{ij}$ then has either the factor s^m or s^{m-1}.

THEOREM 18.3.   [64].   Suppose the compartmental system has $m \geq 1$ traps.   Then

$$q(0_i, I_j) = m$$

if the subsystem of $n(0_i, I_j)$ compartments does not contain a trap.   If this subsystem contains one or more traps, then

$$q(0_i, I_j) = m - 1.$$

Example 2.   To illustrate the theorem, consider the liver disease model reported in Section 16.   In this system, compartments 2 and 4 are traps.   Let

$$I_1 \equiv 1, \; 0_1 \equiv 1, \; 0_2 \equiv 2, \; 0_3 \equiv 4.$$

Then from the connectivity of the system (Figure 16.4), the corresponding shortest path lengths are respectively

$$p(0_1, I_1) = 0, \quad p(0_2, I_1) = 1, \quad p(0_3, I_1) = 2.$$

For the pair $(0_1, I_1)$, the intersection of the set of input reachable compartments with the set of output reachable compartments is $\{1, 2, 3, 4\} \cap \{1, 3\} = \{1, 3\}$. The subsystem $\{1, 3\}$ of $n(0_1, I_1) = 2$ compartments has no traps.   Hence by Theorem 18.3, $q(0_1, I_1) = m = 2$.   Likewise for the pairs $(0_2, I_1)$ and $(0_3, I_1)$ we have

$$\text{subsystem} \; \{1, 2, 3\} \qquad q(0_2, I_1) = m - 1 = 1$$
$$\text{subsystem} \; \{1, 3, 4\} \qquad q(0_3, I_1) = m - 1 = 1$$

since each of these submodels contains a trap.   This analysis shows that the last $m = 2$ coefficients in the fourth degree characteristic polynomial $\det(sI - A)$ will be zero.   Moreover, the first $p(0_i, I_j)$ coefficients and the last $q(0_i, I_j)$ coefficients in the numerator of $\hat{\phi}_{ij}$ will be zero.   All of these facts are confirmed if we actually calculate the 3×1 matrix $\hat{\phi}$:

$$(18.13) \quad \begin{bmatrix} 1 \ s^3 - a_{33} \ s^2 + 0 \ s + 0 \\ 0 \ s^3 + a_{21} \ s^2 - a_{21} \ a_{33} \ s + 0 \\ 0 \ s^3 + 0 \quad s^2 + a_{43} \ a_{31} \ s + 0 \end{bmatrix} \Big/ \Delta(s)$$

where the characteristic polynomial

$$\Delta(s) = s^4 - (a_{11} + a_{33})s^3 + (a_{11}a_{33} - a_{13}a_{31})s^2 + 0 \ s + 0 .$$

As a consequence of Theorem 18.3, formula (18.4) is further refined (in the case of a system with m > 1 traps) to

$$(18.14) \quad \sum_{i=1}^{p} \sum_{j=1}^{q} [n - p(0_i, I_j) - q(0_i, I_j)] + (n - m)$$

which gives the maximal number of nonzero coefficients present in $\hat{\phi}$. For the preceeding example, expression (18.14) computes to

$$[4 - 1 - 2] + [4 - 1 - 1] + [4 - 2 - 1] + (4 - 2) = 6.$$

At this point we examine and exclude from formula (18.14) the number of factors, if any, common to the $\hat{\phi}_{ij}$ numerator and the denominator det(sI - A). Let $r(0_i, I_j) \geq 0$ be the degree of the polynomial factor common to this $\hat{\phi}_{ij}$ numerator and the denominator. This number is readily calculated according to the next statement.

THEOREM 18.4 [64]. The integer $r(0_i, I_j)$ is equal to the residual $n - n(0_i, I_j)$.

Armed with this knowledge, we see that the quantity $r(0_i, I_j)$ includes $q(0_i, I_j)$. For, if the subsystem of $n(0_i, I_j)$ compartments contains no traps, then $n(0_i, I_j)$ can be no larger than n - m, and so by Theorems 18.4 and 18.3,

$$r(0_i, I_j) = n - n(0_i, I_j) \geq m = q(0_i, I_j).$$

Now suppose the subsystem contains one or more traps. For the moment we assume the m traps

(18.15) $\{T_1, T_2, \ldots, T_m\}$

of the entire compartmental system are singletons which do not include compartment O_i. Since the set R of O_i - reachable compartments cannot have set (18.15) as a subset, then

$$n(O_i, I_j) \leq \text{card } R \leq n - \text{card } \{T_1, \ldots, T_m\} = n - m,$$

where card refers to the number of compartments in the respective sets. If, on the other hand, O_i is one of the traps, then

$$n(O_i, I_j) \leq n - \text{card } \{T_k\} = n - (m - 1).$$

If any of the traps in (18.15) are made up of more than single compartments, then $n(O_i, I_j)$ is smaller still. Hence all cases considered we have

$$n(O_i, I_j) \leq n - m + 1.$$

Thus

$$r(O_i, I_j) = n - n(O_i, I_j) \geq m - 1 = q(O_i, I_j)$$

for the subsystem containing traps.

Strict inequality can occur between $r(O_i, I_j)$ and $q(O_i, I_j)$. For example, consider Figure 18.1 with $I_1 = 1 = O_1$. Then the subsystem of $n(O_1, I_1)$ compartments is $\{1, 2, 3\} \cap \{1\} = \{1\}$, a submodel with no trap in a system with $m = 1$ trap. Hence

$$r(O_1, I_1) = n - n(O_1, I_1) = 3 - 1 > m = q(O_1, I_1).$$

As a consequence of the fact that $r(O_i, I_j)$ includes $q(O_i, I_j)$, formula (18.14) is improved to

(18.16) $$N \equiv \sum_{i=1}^{p} \sum_{j=1}^{q} [n(O_i, I_j) - p(O_i, I_j)] + (n - m)$$

wherein $r(O_i, I_j)$ replaces $q(O_i, I_j)$ in (18.14). Expression (18.16) counts the maximal number of coefficients in $\hat{\phi}$ not excluded either because they are nonzero or

because they are not involved in factors common to both the numerator and denominator of $\hat{\phi}_{ij}$.

As other causes of dependency can exist among the coefficients of $\hat{\phi}$ besides
those considered in leading up to (18.16), the number ν of independent equations
arising from the transfer function is less than or equal to N.

Example 2 (continued). We again consider the liver disease model whose connectivity is given in Figure 16.4. As was observed a few pages back,

$$I_1 = 1, \quad O_1 = 1, \quad O_2 = 2, \quad O_3 = 4,$$

(18.17) $\quad p(O_1, I_1) = 0, \quad p(O_2, I_1) = 1, \quad p(O_3, I_1) = 2,$

$$n(O_1, I_1) = 2, \quad n(O_2, I_1) = 3, \quad n(O_3, I_1) = 3.$$

Thus the double sum of (18.16) becomes

$$N = [2-1] + [3-1] + [3-2] + (4-2) = 6,$$

which agrees with the count of nonzero coefficients involving a_{ij} that can be made
from (18.13). Hence $\nu \leq 6$. Reading from (18.13), the 6 equations are

$$\gamma_1 = -a_{33} = a_{13} + a_{43}$$

$$\gamma_2 = a_{21}$$

(18.18) $\quad \gamma_3 = -a_{21}a_{33} = a_{21}(a_{13} + a_{43})$

$$\gamma_4 = a_{43}a_{31}$$

$$\gamma_5 = -a_{11} - a_{33} = a_{21} + a_{31} + a_{13} + a_{43}$$

$$\gamma_6 = a_{11}a_{33} - a_{31}a_{13} = a_{21}a_{13} + a_{21}a_{43} + a_{31}a_{43}$$

put in terms of the four positive parameters a_{21}, a_{13}, a_{31}, and a_{43}. The third and
sixth equations in (18.18) are dependent since

$$\gamma_3 = \gamma_1\gamma_2 \qquad\qquad \gamma_6 = \gamma_1\gamma_2 + \gamma_4.$$

The remaining four equations have Jacobian matrix

$$J \equiv \begin{bmatrix} 0 & 1 & 0 & 1 \\ 1 & 0 & 0 & 0 \\ 0 & 0 & a_{43} & a_{31} \\ 1 & 1 & 1 & 1 \end{bmatrix}$$

whose rank is 4 since $\det J = a_{31}$, which is positive at any admissible value a_{31}.

Therefore $\nu = 4$. These independent equations uniquely determine the four transfer model parameters:

$$\overset{*}{a}_{21} = \gamma_2$$

$$\overset{*}{a}_{31} = \gamma_5 - \gamma_2 - \gamma_1$$

$$\overset{*}{a}_{43} = \gamma_4 / (\gamma_5 - \gamma_2 - \gamma_1)$$

$$\overset{*}{a}_{13} = \gamma_1 - \gamma_4 / (\gamma_5 - \gamma_2 - \gamma_1).$$

So long as these values for $\overset{*}{a}_{ij}$ turn out to be positive, this tracer model is (uniquely) structurally identifiable from the particular experimental setup.

Exercise 18.6.   Compute the value N of formula (18.16) for the compartmental system in Figure 18.1 in which compartment 1 is input and compartments 1 and 2 are sampled. Calculate $\hat{\phi}(s)$, simplify it, and compare the number of its nonzero coefficients with N.

18D.   The Nonidentifiability Test

The real value of having the readily computable expression (18.16) is that it leads immediately to a test concerning nonidentifiability that is referred to in the title of this section. We wish to have a priori evaluation of whether knowledge of all the coefficients in the reduced form of $\hat{\phi}$ --that is, all information available about the model behavior for the given experimental setup--allows determination of the unknown transfer parameters a_{ij} in A. The number ν of independent equations involving the a_{ij} is to be compared with the number μ of unknown a_{ij}. If $\nu = \mu$, then solutions for the a_{ij} can exist but may not be unique. If $\nu < \mu$, then the model is probably structurally nonidentifiable since a nonlinear algebraic system of equations with more unknowns than equations usually does not have a unique solution. Since we now know from (18.16) that $\nu \leq N$, then the condition

$$(18.19) \qquad \boxed{N < \mu}$$

constitutes a simple test for nonidentifiability of a model for a preselected experiment.

Example 2 (continued). With the purpose of illustrating this nonidentifiability test, we again treat the four compartment liver disease model, for which there are $\mu = 4$ unknown a_{ij} transfer parameters. However, this time a complete identifiability analysis is presented with reference to Table 18.1 in which an increasingly complex series of possible tracer experiments is listed. We intend to use the non-identifiability test to indicate the feasible experiments among those listed in the table. The biologist might say do Experiment III since it is actually possible to sample compartments 1, 2, and 4, and therefore get the maximum amount of data. In fact, Experiment III is the experiment that was performed in this study [199], [9]. By contrast, the biomathematician might be tempted to say that the experimentalist should carry out the least complicated tracer study for which structural identification is possible, all other things being equal. As will be seen presently, that study is Experiment II, not III.

TABLE 18.1.

Possible Liver Disease Model Identification Experiments

Experiment	Input Compartment	Output Compartments
I	1	1
II	1	1, 2
III	1	1, 2, 4

In Experiment I, compartment 1 is bolus input and subsequently sampled, so that we select

$$B^T \equiv [1 \quad 0 \quad 0 \quad 0] \equiv C$$

in the system (A, B, C). Using the calculations displayed in (18.17), it follows from (18.16) that

$$N = [2-1] + (4-2) = 3 < \mu$$

so that condition (18.19) holds. Hence we could not in good faith recommend Experiment I to be performed as an identification study by the bioscientist because there will be fewer equations than unknowns a_{ij}.

Exercise 18.7.   Show that, for Experiment I, $\hat{\phi}$ is the first row of (18.13) and thus there will be only three equations in the four unknowns from which a unique solution for the a_{ij} is not available.

This brings us to Experiment II, where the model is system (A, B, C) with

$$B^T \equiv [1 \quad 0 \quad 0 \quad 0] \qquad C \equiv \begin{bmatrix} 1 & 0 & 0 & 0 \\ 0 & 1 & 0 & 0 \end{bmatrix}.$$

Again from (18.17) we get the ingredients to calculate formula (18.16):

$$N = [2-1] + [3-1] + (4-2) = 5.$$

Because condition (18.19) does not hold, Experiment II is a feasible experiment.

Exercise 18.8.   For Experiment II, derive the N = 5 equations in the a_{ij}. Show that of these there is one dependent equation. From the remaining four equations, find the unique solution for the four a_{ij}.

Experiment III was treated above in (18.17) and (18.18). We found N = 6 > μ.

The overall results of our analysis are presented in Table 18.2. Condition (18.19) is satisfied for the simplest experiment. The least complicated possibly successful experiment in the list is II. It is verified through Exercise 18.8 that

Experiment II is in fact a successful identification study. Also Experiment III is successful. This example demonstrates the usefulness of (18.19) in eliminating in advance, directly from information easily obtainable from the connectivity diagram of the model, probable unsuccessful identification experiments.

TABLE 18.2.

Summary of Nonidentifiability Test for the
4-Compartment Liver Disease Model

Experiment	Feasible Experiment
I	NO
II	YES
III	YES

The treatment thus far in this section has dealt with compartments receiving individual inputs and being individually sampled. This is a very common type of tracer experiment. However, we may extend our considerations to tracer studies in which more than one compartment receives the same input u_j and multiple compartments are involved in the same output y_i. In such cases, $p(0_i, I_j)$ is extended to the quantity $p(y_i, u_j)$ which is the shortest path length in the connectivity diagram of the model from one of the compartments receiving input u_j to one of the compartments directly being measured in output function component y_i. Analogous interpretations hold for $n(y_i, u_j)$, $q(y_i, u_j)$, and $r(y_i, u_j)$. Expression (18.16) for N is replaced by

(18.20)
$$N \equiv \sum_{i=1}^{p} \sum_{j=1}^{q} [n(y_i, u_j) - p(y_i, u_j)] + (n - m).$$

Cobelli et al. [64], [61] present a further extension of this formula.

18E. Tighter Bounds on the Number of Independent Equations

One case of special interest where more than one compartment is involved in the same output is when the only measurement of the system available is the total amount of tracer present:

$$y(t) = Cx(t) \equiv [1 \quad 1 \quad \dots \quad 1]x(t) = \sum_{i=1}^{n} x_i(t).$$

This single output function is called the <u>washout function</u> or the whole body retention curve and it is often obtained by administrating tracer to the tracee input of a steady state system at a constant level until a steady state of tracer is achieved, and then terminating the tracer input and following the total activity of the system [134, p. 50], [34].

<u>Exercise 18.9.</u>   Investigate the <u>convexity</u> of the washout curve [122]. What is the relationship between the washout function and the distribution function of exit times in Section 14?

When the output of a system is the washout curve, we have the following result.

<u>THEOREM 18.5.</u>   Suppose that in the compartmental system (A, B, C), the matrix $C \equiv [1 \quad 1 \quad \dots \quad 1]$ so that the output y is the washout function. Moreover, suppose no compartment receiving input has an excretion to the outside environment. Then the number N in (18.20) is reduced to

$$(18.21) \quad N - q = \sum_{j=1}^{q} n(y_1, u_j) - 2q + n - m.$$

Proof.   The first two coefficients in the numerator of $\hat{\phi}$ are

$$\Gamma_1 \equiv M_0$$

$$\Gamma_2 \equiv M_1 + s_1 M_0$$

from (17.2). Since C is a $1 \times n$ matrix, and B is $n \times q$, then the matrices Γ_i are of dimension $1 \times q$. In particular,

$$\Gamma_1 = CB \equiv [\gamma_1 \ \gamma_2 \ \dots \ \gamma_q],$$

where the generic element γ_j is the sum of the jth column of B. Each of the numbers γ_j is nonzero by the definition of the B-matrix. Since no compartment directly receiving input has excretion to the outside, the matrix AB has all column sums equal to zero. Hence the Markov parameter $M_1 = CAB = 0$. Thus

$$(18.22) \qquad \Gamma_2 = M_1 + s_1 M_0 = s_1 [\gamma_1 \ \gamma_2 \ \cdots \ \gamma_q],$$

and so we see that every equation in the a_{ij} introduced through the coefficient Γ_2 is just a nonzero constant multiple of s_1, the scalar coefficient available from the denominator in (17.2). Since the trace of A must be nonzero, then $s_1 \neq 0$ and hence each entry in Γ_2 is nonzero. Thus s_1 as well as the elements in Γ_2 are counted in the number N of coefficients available from $\hat{\phi}$ in (18.20). However, it is now clear from (18.22) that each of the q entries in Γ_2 is just a multiple of s_1. Therefore the number ν of independent equations from the transfer function for the a_{ij} can be no greater than N - q. Since every compartment is involved in the output, we must have p = 1 and $p(y_1, u_j) = 0$ for all j. Hence expression (18.20) is replaced by

$$\sum_{j=1}^{q} [n(y_1, u_j) - 1] + (n - m) - q = \sum_{j=1}^{q} n(y_1, u_j) - 2q + n - m.$$

▨

To illustrate the above theorem, we refer to Teorell's system (A, B, C) of Figure 16.5 in which we take

$$B \equiv [1 \quad 0 \quad 0]^T \qquad\qquad C \equiv [1 \quad 1 \quad 1].$$

The compartmental matrix

$$A \equiv \begin{bmatrix} a_{11} & 0 & 0 \\ a_{21} & a_{22} & a_{23} \\ 0 & a_{32} & a_{33} \end{bmatrix} \qquad \begin{aligned} a_{11} &= -a_{21} \\ a_{22} &= -a_{32} - a_{02} \\ a_{33} &= -a_{23} - a_{03} \end{aligned}$$

has $\mu = 5$ unknown model parameters a_{ij} $(i \neq j)$. Compartment 1 is tracer input but has no outside excretion, so that Theorem 18.5 applies. The set of u_1 - reachable compartments as well as the set of y_1 - reachable compartments is {1, 2, 3}. Hence

$n(y_1, u_1) = 3$. Now from (18.20) there obtains

$$N = [3 - 1] + (3 - 0) = 5.$$

According to condition (18.19), this experiment would be accepted as feasible. However, the more refined information of Theorem 18.5 yields

$$N - q = 3 - 2 \cdot 1 + 3 - 0 = 4$$

from (18.21), so that now the new condition replacing (18.19), $N - q < \mu$, is met and strongly suggests rejecting this experiment. To check this, the five coefficients of $\hat{\phi}$ which involve a_{ij} are calculated to be

$$s_1 = a_{21} + a_{32} + a_{02} + a_{23} + a_{03}$$

$$s_2 = a_{21}a_{32} + a_{21}a_{02} + a_{21}a_{23} + a_{21}a_{03} + a_{32}a_{03} + a_{02}a_{23} + a_{02}a_{03}$$

(18.23) $\quad s_3 = a_{21}a_{32}a_{03} + a_{21}a_{02}a_{23} + a_{21}a_{02}a_{03}$

$$M_1 + s_1 M_0 = 0 + s_1 \cdot 1$$

$$M_2 + s_1 M_1 + s_2 M_0 = a_{21}a_{23} + a_{21}a_{03} + a_{32}a_{03} + a_{02}a_{23} + a_{02}a_{03}$$

in which we see the duplication of the first and fourth coefficients. Thus there are $N - q = 4$ equations in the $\mu = 5$ unknowns.

Exercise 18.10.   Demonstrate Theorem 18.5 on the system (A, B, C) of Exercise 16.7. Compute N of (18.20), N - q of (18.21), and ν.

Exercise 18.11.   Suppose A is a $n \times n$ SEC mammillary system in which the mother compartment $\equiv$ number 1 $\equiv$ exit compartment. Also suppose a_{1j}, $a_{j1} \neq 0$ for all $j = 2, 3, \ldots, n$. Let B be the jth column of the $n \times n$ identity matrix for any $j \geq 2$, and C $\equiv$ [1   1 $\ldots$ 1]. Discuss the nonidentifiability test based on (18.20) and then on (18.21). What happens if, say $a_{1n} = a_{1,n-1}$? Under this condition can explicit equations be derived for the a_{j1} and a_{1j} (see [34, p. 132])?

Theorem 18.5 presents a tightening of the upper bound N in the inequality $\nu \leq N$. Numerous other results of the same flavor, which involve detecting dependency among the equations arising from the coefficients of the transfer function, are given by Anderson [6]. More research needs to be done in refining these bounds on the number of independent nonlinear equations in the model parameters a_{ij}.

SECTION 19.   COMPUTATION OF THE MODEL PARAMETERS

19A. Local Identifiability

We consider now the problem of computing an isolated solution for the model parameters $\{a_{ij}\}$, $i \neq j$, from the nonlinear algebraic system of equations (17.4) and (17.5). This structural identification of $A \equiv [a_{ij}]$ is difficult to treat from a systematic theoretical viewpoint since it requires the inversion of a system of non-linear algebraic equations [177] (also see Section 22). If an explicit solution of a system of nonlinear equations cannot be displayed, then such problems are usually solved by iteration. Probably the best known method for attacking this problem is the Contractor Mapping Principle, which allows us to show existence and uniqueness of isolated solutions. In particular, Newton's method--under certain restrictions-- acts locally about a solution as a contractor, and via the Kantorovich conditions there can be established the existence of a solution unique to within a certain neighborhood.

Let us assume that necessary conditions for the structural identifiability of A are satisfied, and that the number ν of independent equations in (17.4) and (17.5) arising from the coefficients of $\hat{\phi}$ is equal to the number μ of unknown model param- eters a_{ij}, $i \neq j$. A procedure such as used in the example of (18.18) for deter- mining the number of independent equations may be adopted. Even though $\nu = \mu$, globally unique identification of A may not occur. For example, consider the fol- lowing system (A, B, C), which is also mentioned by Delforge [82]:

$$A \equiv \begin{bmatrix} a_{11} & 0 \\ a_{21} & a_{22} \end{bmatrix} \qquad \begin{aligned} a_{11} &= - a_{21} \\ a_{22} &= - a_{02} \end{aligned}$$

$$B \equiv [1 \quad 0]^T \qquad\qquad C \equiv [1 \quad 1]$$

(see Figure 17.2). The transfer function for this system is

$$\hat{\phi}(s) = (s + \gamma_1)/(s^2 + \gamma_1 s + \gamma_2)$$

where

$$\gamma_1 = a_{21} + a_{02}$$

(19.1)

$$\gamma_2 = a_{21}a_{02} \cdot$$

Note the duplication in γ_1 between the numerator and denominator of $\hat{\phi}$ (as predicted by Theorem 18.5). The number of independent coefficients in $\hat{\phi}$ is

$$\nu = N - q = 3 - 1 = 2,$$

which is the same as the number $\mu = 2$ of unknown model parameters. The complete symmetry between a_{21} and a_{02} in (19.1) shows that there are two (and only two) isolated solutions of this system. For instance, if $\gamma_1 = 3$ and $\gamma_2 = 2$, then there are solutions

$$\alpha^* = (a_{21}^*, a_{02}^*) = (2, 1); \quad \alpha^{**} = (a_{21}^{**}, a_{02}^{**}) = (1, 2).$$

We say that the above model is "locally identifiable" at α^* and also at α^{**}. Let $y(t,\alpha)$ be the system (A, B, C) output function, $\bar{y}$ the data vector collected over the interval [0, T], and

$$V(\alpha) \equiv \int_0^T \| y(t, \alpha) - \bar{y}(t) \|_2^2 \, dt$$

(the integral may be replaced by a sum in the discrete time case). Bellman and Astrom [24] say the model (A, B, C) is <u>locally identifiable at $\hat{\alpha}$</u> (in the least squares sense) provided $V(\alpha)$ has a local minimum at $\hat{\alpha}$ in parameter space. It can be shown that a sufficient condition for local identifiability at $\hat{\alpha}$ is the rank of the Jacobian of the nonlinear algebraic system evaluated at $\hat{\alpha}$ is equal to the number of independent parameters [109].

19B. <u>Newton's Method and Modifications</u>

Equations (17.4) and (17.5) from the transfer function normally give rise to a system of ν independent equations in $\nu = \mu$ unknowns

$$F_i(\alpha_1, \alpha_2, \ldots, \alpha_\nu) = 0, \qquad 1 \le i \le \nu,$$

which are the component equations of $F(\alpha) = 0$. Here the α_j represent individual $a_{k\ell}$ ($k \neq \ell$) and α is the vector of the α_j. Due to the particular structure of equations (17.4) and (17.5), the typical function F_i will consist of a sum of terms, each term being the product of certain distinct unknowns among α_1, α_2, ..., α_ν (see Section 22). Equations (18.18) and (18.23) illustrate this form. Thus F is continuously differentiable in R^ν. Let $F'(\alpha)$ denote the Jacobian matrix $[(\partial F_i/\partial\alpha_j)(\alpha)]$ evaluated at α. As the Jacobian is readily obtainable in this problem, Newton's method [177, p. 312]

(19.2) $\qquad \alpha^{k+1} \equiv \alpha^k - F'(\alpha^k)^{-1} F(\alpha^k), \qquad\qquad k = 0, 1, \ldots,$

is very attractive as a possible algorithm for the solution of $F(\alpha) = 0$. The method takes a starting point α^0 (which is thought to be a reasonable approximation of a solution α^* of $F(\alpha) = 0$) and attempts to improve on α^0 by iteration (19.2).

As an example, we take the system (A, B, C) where A is the 3×3 catenary matrix

$$
A \equiv \begin{bmatrix} a_{11} & a_{12} & 0 \\ a_{21} & a_{22} & a_{23} \\ 0 & a_{32} & a_{33} \end{bmatrix}
\qquad
\begin{aligned}
a_{11} &= -a_{21} - a_{01} \\
a_{22} &= -a_{12} - a_{32} \\
a_{33} &= -a_{23}
\end{aligned}
$$

for a single exit system with excretion only from compartment 1 [8]. Moreover, let $B^T \equiv [1 \quad 0 \quad 0] \equiv C$. Via (17.2) we compute

$$\hat{\phi}(s) = (s^2 + \gamma_1 s + \gamma_2)/(s^3 + \gamma_3 s^2 + \gamma_4 s + \gamma_5)$$

where

$$a_{12} + a_{32} + a_{23} - \gamma_1 \equiv F_1 = 0$$

$$a_{12}a_{23} - \gamma_2 \equiv F_2 = 0$$

(19.3) $\quad a_{21} + a_{01} + a_{12} + a_{32} + a_{23} - \gamma_3 \equiv F_3 = 0$

$$a_{01}a_{12} + a_{21}a_{32} + a_{01}a_{32} + a_{21}a_{23} + a_{01}a_{23} + a_{12}a_{23} - \gamma_4 \equiv F_4 = 0$$

$$a_{01}a_{12}a_{23} - \gamma_5 \equiv F_5 = 0.$$

Exercise 19.1.   In (19.3) suppose the γ-vector is [0.78, 0.015, 1.25, 0.31, 0.0025] (more in Section 23 on how we arrive at the γ_i parameter estimates). Use the following steps for k = 0, 1, 2, ..., to find a positive solution of (19.3):

(19.4)

      (a)   compute $F(\alpha^k)$;

      (b)   if α^k is acceptable, stop; otherwise compute $F'(\alpha^k)$;

      (c)   solve the linear system $F'(\alpha^k)\Delta^k = - F(\alpha^k)$ for the vector Δ^k;

      (d)   set $\alpha^{k+1} \equiv \alpha^k + \Delta^k$.

Experiment with different starting values. How are these initial values chosen?

Exercise 19.2.   For systems $F(\alpha) = 0$ associated with compartmental problems arising from (17.4) and (17.5), why must the Jacobian $F'(\alpha)$ always have one row which consists entirely of ones?

There are variations of Newton's method so as to improve its computational efficiency. One popular technique for trying to reduce the effort involved in the algorithm is to hold the Jacobian fixed for all of the iterations. If, in (19.2) or (19.4), the derivative $F'(\alpha^k)$ is replaced by $F'(\alpha^0)$ for all k = 0, 1, 2, ..., the iteration is known as the simplified Newton method or the chord method [177, p. 312], [176, p. 151]. In general this process will decrease the rate of convergence (if convergence takes place), but it may be more efficient overall. The usual method of handling part (c) of (19.4) is to form the L(k)U(k) - decomposition of $F'(\alpha^k)$ [222], where L(k) is a lower triangular matrix and U(k) is an upper triangular matrix, so that forward and then backward substitution can be used in solving the linear system for Δ^k. If k is fixed at zero in just the Jacobian, then the advantage gained is that the factorization $F'(\alpha^0)$ = LU can be retained throughout the entire iteration sequence.

There is another modification of Newton's method that is suggested directly from the particular type of problem that arises from compartmental studies. Perhaps the use of this algorithm is best motivated through an example. Let us rearrange and redefine the five equations of (19.3) and Exercise 19.1 as

$$\alpha_1 + \alpha_2 + \alpha_3 + \alpha_4 + \alpha_5 - 1.25 = 0$$

$$\alpha_2\alpha_4 + \alpha_1\alpha_3 + \alpha_2\alpha_3 + \alpha_1\alpha_5 + \alpha_2\alpha_5 + \alpha_4\alpha_5 - 0.31 = 0$$

(19.5) $\qquad \alpha_3 + \alpha_4 + \alpha_5 - 0.78 = 0$

$$\alpha_2\alpha_4\alpha_5 - 0.0025 = 0$$

$$\alpha_4\alpha_5 - 0.015 = 0$$

where

$$\alpha_1 \equiv a_{21}, \ \alpha_2 \equiv a_{01}, \ \alpha_3 \equiv a_{32}, \ \alpha_4 \equiv a_{12}, \ \alpha_5 \equiv a_{23} \ .$$

The rationale for this particular rearrangement of variables and equations is that
the Jacobian

$$(19.6) \qquad F'(\alpha) = \begin{bmatrix} 1 & 1 & 1 & 1 & 1 \\ \alpha_3+\alpha_5 & \alpha_3+\alpha_4+\alpha_5 & \alpha_2 & \alpha_2+\alpha_5 & \alpha_1+\alpha_2+\alpha_4 \\ 0 & 0 & 1 & 1 & 1 \\ 0 & \alpha_4\alpha_5 & 0 & \alpha_2\alpha_5 & \alpha_2\alpha_4 \\ 0 & 0 & 0 & \alpha_5 & \alpha_4 \end{bmatrix}$$

turns out to be <u>nearly upper triangular</u>. Moreover, the nonzero entries below the
diagonal ($\alpha_3 + \alpha_5$, $\alpha_4\alpha_5$, and α_5) tend to be small in the region $0 < \alpha_j < 1$ for which
there is a positive solution to this problem. Hence another modification of Newton's
method, which is suggested by the compartmental analysis, is the iterative scheme

(19.7) $\qquad \alpha^{k+1} \equiv \alpha^k - U(\alpha^k)^{-1}F(\alpha^k) \qquad\qquad k = 0, 1, \ldots,$

where $U(\alpha)$ is defined as the upper triangular part of $F'(\alpha)$, including the diagonal.
This algorithm reduces the number of partial derivative calculations required, and
the inversion $U(\alpha^k)^{-1}F(\alpha^k)$ is easily accomplished by backward substitution at each
step of the process.

Exercise 19.3.   Redo Exercise 19.1 using scheme (19.7).

It should be noted that there are cases in compartmental models where $F'(\alpha)$ is exactly upper triangular. Consider the model of Figure 16.1 ($a_{02} \equiv 0$) associated with, for example, lipoprotein synthesis [37]. The 3×3 system for the identification problem is

$$0 = \alpha_1 + \alpha_2 + \alpha_3 - \gamma_1$$

(19.8) $$0 = \alpha_2 \alpha_3 - \gamma_2$$

$$0 = \alpha_3 - \gamma_3$$

where $\alpha_1 \equiv a_{21}$, $\alpha_2 \equiv a_{01}$, and $\alpha_3 \equiv a_{12}$.

To study more closely the feasibility of iteration (19.7), we introduce for the function F the auxiliary function $G : D \to R^\nu$ by

$$G(\alpha) \equiv \alpha - U(\alpha)^{-1} F(\alpha)$$

defined on the domain $D \subseteq R^\nu$ of points α for which $U(\alpha)$ is invertible. Upon differentiating G at a solution α^* of $F(\alpha) = 0$, we get

(19.9) $$G'(\alpha^*) = I - U(\alpha^*)^{-1} F'(\alpha^*) - [U(\alpha^*)^{-1}]' F(\alpha^*) .$$

Let $L(\alpha)$ be the strictly lower triangular part of $F'(\alpha)$ so that $F'(\alpha) = U(\alpha) + L(\alpha)$. Then, since $F(\alpha^*) = 0$, equation (19.9) becomes

$$G'(\alpha^*) = - U(\alpha^*)^{-1} L(\alpha^*).$$

Hence, by Ostrowski's Theorem [176, p. 145] and the Contractor Mapping Principle, we require the spectral radius of the matrix $U(\alpha^*)^{-1} L(\alpha^*)$ to be less than unity in order to have α^* as a point of attraction of the iteration $\alpha^{k+1} \equiv G(\alpha^k)$. We can argue heuristically that there is a good chance that this sufficient condition will hold since, due to the rearrangement of equations motivating the design of algorithm (19.7), the matrix $L(\alpha^*)$ may contain enough zeros below the diagonal to ensure

$$\| U(\alpha^*)^{-1} L(\alpha^*) \| < 1$$

for some norm $\| \cdot \|$ on R^ν.

An example using Newton's Method for the determination of rate constants in a mammillary system of a steady state tracer study is given in [117].

19C. The Kantorovich Conditions

Let us suppose our function F is as described above, and by some scheme (such as the simplified Newton method, Newton's method, iteration (19.7), or say, sub-routine ZSCNT or ZSYSTM in the IMSL package [131]), we have arrived at a value which we call α^0, for which $\| F(\alpha^0) \|$ is small. Then further use of either the simplified Newton method or Newton's method [177, p. 421], [40] will indicate the local identifiability of the model and in fact will produce a uniqueness neighborhood affiliated with this isolated solution. This is done by applying the Kantorovich conditions (a) - (d) below to the function F at α^0.

THEOREM 19.1.   [177], [40]. Suppose there is an α^0 in a convex set D_0 and a norm $\| \cdot \|$ such that

(a)   for all α, β in D_0,

(19.10) $\| F'(\alpha) - F'(\beta) \| \le \kappa \| \alpha - \beta \|$

for some constant κ;

(b)   the matrix $F'(\alpha^0)$ is invertible and

$\| F'(\alpha^0)^{-1} \| \le b$;

(c)   the norm $\| F'(\alpha^0)^{-1} F(\alpha^0) \| \le n$;

(d)   the product $\kappa b n \le 1/2$.

Set

$$t^* \equiv (b\kappa)^{-1} [1 - (1 - 2\kappa b n)^{1/2}] \qquad t^{**} \equiv (b\kappa)^{-1} [1 + (1 - 2\kappa b n)^{1/2}]$$

and assume that the closed sphere

$$\bar{S}(\alpha^0, t^*) \equiv \{z \in R^\nu : \| z - \alpha^0 \| \leq t^* \}$$

is a subset of D_0. Then the simplified Newton iterates or the Newton iterates starting from α^0 are well-defined, remain in $\bar{S}(\alpha^0, t^*)$, and there exists a unique solution α^* of $F(\alpha) = 0$ in the neighborhood

(19.11) $\{z \in R^\nu : \| z - \alpha^0 \| < t^{**} \} \cap D_0$

to which either of these iterates converge.

The proof of the Theorem involves applying the Contractor Mapping Principle to the function

$$H(\alpha) \equiv \alpha - F'(\alpha)^{-1} F(\alpha)$$

in some subdomain of R^ν. Also an investigation is made of upper bounds for $\| H'(\alpha) \|$. The combination $\kappa b \eta$ is a contractor constant, and the essence of Theorem 19.1 is that if $\kappa b \eta \leq 1/2$ at $\alpha = \alpha^0$, then $\kappa b \eta$ remains less than unity for all future iterates and thus convergence takes place to a unique solution α^* of $F(\alpha) = 0$ in (19.11).

It seems that the best way to use Theorem 19.1 is to start computing iterates by whatever method until it appears that $\| \alpha^j - \alpha^{j-1} \| < \epsilon$ for prescribed ϵ, and $\| F(\alpha^j) \|$ is small. Then redefine α^j as the guess α^0 and use the Theorem on this α^0 to prove existence and uniqueness of a solution α^* of $F(\alpha) = 0$ in the region (19.11).

Probably the most difficult of the above conditions to verify is the Lipschitz condition (a) for F'. Due to the form of F arising from the transfer function, the ∞-norm is recommended for use in the Theorem. This is because of the type of expressions present in $F'(\alpha) - F'(\beta)$ of (19.10). As an example, this matrix may involve entries such as

(19.12) $\alpha_j \alpha_k - \beta_j \beta_k$

(see (19.6) for instance). Such differences can be handled by the following trick in order to establish the Lipschitz condition (19.10): add and subtract $\beta_j \alpha_k$ but regroup terms. Thus (19.12) is equal to

$$(\alpha_j \alpha_k - \beta_j \alpha_k) + (\beta_j \alpha_k - \beta_j \beta_k) = (\alpha_j - \beta_j)\alpha_k + (\alpha_k - \beta_k)\beta_j$$

and so the absolute value of (19.12) is less than or equal to

(19.13) $|\alpha_j - \beta_j| \cdot |\alpha_k| + |\alpha_k - \beta_k| \cdot |\beta_j|$.

Now if D_0 is preselected as the set

$$\{z \in R^\nu : \|z\|_\infty \le c\}$$

for some constant c, then (19.13) is less than or equal to

(19.14) $|\alpha_j - \beta_j|c + |\alpha_k - \beta_k|c$.

Each of these terms is in turn dominated by $c\|\alpha - \beta\|_\infty$. Higher ordered terms like $\alpha_i \alpha_j \alpha_k - \beta_i \beta_j \beta_k$ can be handled similarly: add and subtract $\beta_i \alpha_j \alpha_k$ so that the analysis is reduced to dealing with quadratic expressions like (19.12), which can then be treated as in (19.13) - (19.14).

19D. An Example

For the purpose of illustrating Theorem 19.1, consider the nonlinear system

$$0 = F_1(\alpha) \equiv \alpha_1 + \alpha_2 + \alpha_3 - 1.69$$

(19.15) $$0 = F_2(\alpha) \equiv \alpha_1 \alpha_2 + \alpha_2 \alpha_3 - 0.713$$

$$0 = F_3(\alpha) \equiv \alpha_1 \alpha_2 \alpha_3 - 0.121.$$

Then

$$F'(\alpha) = \begin{bmatrix} 1 & 1 & 1 \\ \alpha_2 & \alpha_1 + \alpha_3 & \alpha_2 \\ \alpha_2 \alpha_3 & \alpha_1 \alpha_3 & \alpha_1 \alpha_2 \end{bmatrix} .$$

Suppose that from the physical limitations of the problem, we know that each model parameter α_i satisfies $0 < \alpha_i \leq 1$. Thus we introduce

$$D_0 \equiv \{(z_1, z_2, z_3) \ \varepsilon \ R^3 : \ 0 < z_i \leq 1\} \ .$$

We pick a starting point in this set and iterate using the simplified Newton method to arrive at the point $\alpha^0 \equiv [0.647, 0.813, 0.230]^T$ around which the iterates seem to be converging and for which $\| F(\alpha^0) \|_\infty$ is small. Now

$$F'(\alpha^0)^{-1} = \begin{bmatrix} 15.685 & -17.385 & -2.949 \\ -12.703 & 15.625 & 0. \\ -1.982 & 1.760 & 2.949 \end{bmatrix}$$

so that $\| F'(\alpha^0)^{-1} \|_\infty = 36.020 \equiv b$. Moreover,

$$F'(\alpha^0)^{-1} F(\alpha^0) = [3.415 \times 10^{-5}, \ 1.563 \times 10^{-5}, \ -4.977 \times 10^{-5}]^T$$

and so the ∞-norm of this vector is $4.977 \times 10^{-5} \equiv \eta$.

The difference

$$F'(\alpha) - F'(\beta) = \begin{bmatrix} 0 & 0 & 0 \\ \alpha_2 - \beta_2 & (\alpha_1 - \beta_1) + (\alpha_3 - \beta_3) & \alpha_2 - \beta_2 \\ \alpha_2\alpha_3 - \beta_2\beta_3 & \alpha_1\alpha_3 - \beta_1\beta_3 & \alpha_1\alpha_2 - \beta_1\beta_2 \end{bmatrix},$$

and the sum of the absolute values of each of the entries in row two is less than or equal to $4 \cdot \max|\alpha_i - \beta_i|$. Each element in the third row has absolute value

$$\leq |\alpha_j - \beta_j| \cdot 1 + |\alpha_k - \beta_k| \cdot 1$$

by (19.14), for appropriate j and k. Assimilating all this information leads to

$$\| F'(\alpha) - F'(\beta) \|_\infty \leq 6 \| \alpha - \beta \|_\infty$$

for all α, β in D_0. Hence we can select $\kappa \equiv 6$. Then $\kappa b\eta = 0.01076$ so that condition (d) of Theorem 19.1 is met. We calculate $t^* = 5.004 \times 10^{-5}$ and

$t^{**} = 9.204 \times 10^{-3}$. Clearly the closed ball $\overline{S}(\alpha^0, t^*) \subset D_0$. Hence by Theorem 19.1, there exists in the region

$$\| z - \alpha^0 \|_\infty < 9.204 \times 10^{-3} ,$$

a unique solution α^* of the system (19.15). (A more exact approximation of the solution is $[0.64697, 0.81298, 0.23005]^T$.)

Exercise 19.4.   Apply Theorem 19.1 to equations (19.8) in which $\gamma_1 = 2.51$, $\gamma_2 = 0.842$, and $\gamma_3 = 0.818$.

SECTION 20. AN ALTERNATIVE APPROACH TO IDENTIFICATION

20A. A New Identification Method

Different aspects of the identification problem for the linear model (16.5)
of a compartmental system have been discussed in Sections 16-19. These have been
based primarily on the transfer function. In this section we introduce another
technique based on the modal matrix and the related component matrices of A. The
method is fundamentally a blending of the "normal-mode" identifiability analysis
using the modal matrix and its inverse [172], and certain concepts pointed out in
[236] connecting component matrices to the identifiability problem. Also see
Delforge [81], [80].

In previous sections it was observed that if a system (A, B, C) is identifiable at all, then it is identifiable from its impulse response matrix, or equivalently, from its transfer function matrix. Hence identifiability analyses have
usually assumed the impulse response matrix is known. By contrast, the present
approach is designed to test the identifiability of the system based on partial
knowledge of the entries in the component matrices of A.

20B. The Component Matrices of A

Let us consider the linear time-invariant system (A, B, C) where A is an $n \times n$
compartmental matrix that is diagonalizable with eigenvalues λ_i, and where B and C
are known matrices (see [143] for the more general case). Let f be a function
defined on the spectrum of A. Then there are n nonzero $n \times n$ matrices $Z_k \equiv [z_{ij}^{(k)}]$,
called the components of A, such that the spectral resolution holds:

(20.1) $$f(A) = \sum_{k=1}^{n} f(\lambda_k) Z_k .$$

Following [143] we find that the matrices Z_k commute with A, are linearly independent, are independent of f, satisfy the bilinear condition

(20.2) $$Z_k Z_j = \delta_{kj} Z_k \qquad\qquad k, j = 1, 2, \ldots, n,$$

where δ_{kj} is the Kronecker delta, and obey

(20.3) $AZ_k = \lambda_k Z_k$

for all k = 1, 2, ..., n. This last equation demonstrates that column vectors of Z_k different from zero are right eigenvectors of A with the same eigenvalue λ_k. All eigenvectors of A which have the same eigenvalue λ_k are collinear, and hence all the columns of Z_k are collinear resulting in rank $(Z_k) = 1$.

By judiciously choosing functions to be used in representation (20.1), there is a method of determining the component matrices if A is given. To illustrate, suppose n = 3 and select the linearly independent functions

$$f(\lambda) = 1, \lambda - \lambda_1, (\lambda - \lambda_1)(\lambda - \lambda_2),$$

respectively, for use in (20.1). We then get the linear system of component matrices

$$I = Z_1 + Z_2 + Z_3$$

$$A - \lambda_1 I = (\lambda_2 - \lambda_1)Z_2 + (\lambda_3 - \lambda_1)Z_3$$

$$(A - \lambda_1 I)(A - \lambda_2 I) = (\lambda_3 - \lambda_1)(\lambda_3 - \lambda_2)Z_3$$

which can be successively solved for Z_3, Z_2, and Z_1 by back-substitution if λ_1, λ_2, and λ_3 are all distinct.

Exercise 20.1.   For the compartmental matrix

$$A = \begin{bmatrix} -2 & 1 & 0 \\ 2 & -2 & 0 \\ 0 & 1 & 0 \end{bmatrix}$$

verify that $\lambda_1 = -3.414$, $\lambda_2 = -0.586$, $\lambda_3 = 0$, and

$$Z_3 = \begin{bmatrix} 0 & 0 & 0 \\ 0 & 0 & 0 \\ 1 & 1 & 1 \end{bmatrix}$$

$$Z_2 = \begin{bmatrix} 0.500 & 0.354 & 0 \\ 0.707 & 0.500 & 0 \\ -1.207 & -0.854 & 0 \end{bmatrix}$$

$$Z_1 = \begin{bmatrix} 0.500 & -0.354 & 0 \\ -0.707 & 0.500 & 0 \\ 0.207 & -0.146 & 0 \end{bmatrix}.$$

<u>Exercise 20.2.</u>   Prove that if the compartmental matrix A is such that $a_{0j} = 0$ for all j = 1, 2, ..., n, and if $\lambda_k \neq 0$, then

(20.4) $\sum\limits_{i=1}^{n} z_{ij}^{(k)} = 0$ j = 1, 2, ..., n .

<u>Exercise 20.3.</u>   If the component matrices of A have been determined, how can one form A^{-1} for a nonsingular A without explicitly computing the inverse matrix?

   The <u>(column) modal matrix</u> $S \equiv [s_{ij}]$ has been introduced before and we recall that it is the square matrix whose columns are the linearly independent right eigenvectors e_1, ..., e_n corresponding to the eigenvalues λ_1, ..., λ_n of A. The nonzero $n \times 1$ vector f_i is a <u>left eigenvector of A</u> corresponding to λ_i if $f_i^T A = \lambda_i f_i^T$. (These can be computed as right eigenvectors of A^T.) The <u>biorthogonality condition</u> is well known: If $\lambda_i \neq \lambda_j$, then $f_j^T e_i = 0$. If the left eigenvectors are normalized with respect to the right eigenvectors, i.e., $f_i^T e_i = 1$ for all i, then

$$S^{-1} = [f_1^T \ f_2^T \ \cdots \ f_n^T]^T,$$

and thus the diagonalization $A = S \Lambda S^{-1}$, where Λ is the $n \times n$ matrix $[\lambda_i \delta_{ij}]$, is equivalent to the spectral resolution $(f(\lambda) = \lambda$ in (20.1))

$$A = \lambda_1 Z_1 + \lambda_2 Z_2 + \cdots + \lambda_n Z_n.$$

Let S^{-1} have entries $s_{ij}^{(-1)}$. Then the connection between the modal matrix and the component matrices for A is given by

(20.5) $Z_k = e_k f_k^T$ or $z_{ij}^{(k)} = s_{ik} s_{kj}^{(-1)}$

for i, j, k = 1, 2, ..., n.

Exercise 20.4.   Given the compartmental matrix

$$A = \begin{bmatrix} -2. & 1. \\ 1.5 & -1. \end{bmatrix}$$

show that

$$S = \begin{bmatrix} 1. & 1. \\ -0.823 & 1.823 \end{bmatrix}$$

$$Z_1 = \begin{bmatrix} 0.689 & -0.378 \\ -0.567 & 0.311 \end{bmatrix} \qquad Z_2 = \begin{bmatrix} 0.311 & 0.378 \\ 0.567 & 0.689 \end{bmatrix}$$

and verify equation (20.5).

Exercise 20.5.   Prove that if A has distinct eigenvalues, then for any k = 1, 2, ..., n,

(20.6) $\sum_{i=1}^{n} z_{ii}^{(k)} = 1.$

Does (20.6) hold if A is diagonalizable yet has repeated eigenvalues?

Exercise 20.6.   Speculate on other properties of the component matrices when A is an n × n compartmental matrix. In particular, prove that

(20.7) $Z_n \geq 0.$

Can this inequality be refined if in addition A is an irreducible matrix (i.e., its connectivity diagram is strongly connected)?

It is conjectured from the n = 2 case appearing in [236] that due to mass conservation in the compartmental system, diagonal elements of any component matrix of a compartmental matrix are nonnegative:

(20.8) $z_{ii}^{(k)} \geq 0$ $i, k = 1, 2, \ldots, n.$

This condition is illustrated in the above examples. It is already known that

(20.9) $\sum_{k=1}^{n} z_{ij}^{(k)} = \delta_{ij}$ $i, j = 1, 2, \ldots, n.$

from (20.1) with $f(\lambda) \equiv 1$. The same equation yields

(20.10) $A = \lambda_1 Z_1 + \lambda_2 Z_2 + \ldots + \lambda_n Z_n$

via the function $f(\lambda) \equiv \lambda$. Then, if conjecture (20.8) holds, there obtains from
(20.10), $|\lambda_1| \geq |\lambda_2| \geq \ldots \geq |\lambda_n|$, and (20.9),

$$|a_{ii}| \leq \sum_{k=1}^{n} |\lambda_k| \; |z_{ii}^{(k)}|$$

$$\leq |\lambda_1| \sum_{k=1}^{n} |z_{ii}^{(k)}| = |\lambda_1| \sum_{k=1}^{n} z_{ii}^{(k)} = |\lambda_1|$$

for any $i = 1, 2, \ldots, n$. Hence, provided (20.8) is valid,

$$\max_{1 \leq i \leq n} |a_{ii}| \leq |\lambda_1| ,$$

indicating that the maximum magnitude eigenvalue λ_1 of A satisfied condition (12.16)
discussed before.

20C. The Identification Technique Using Component Matrices

Let us now move on to the identification problem. The condition $AS = S\Lambda$ tells
us that provided we can determine the modal matrix S from observations, then
together with the measured eigenvalues in Λ, these matrices uniquely identify the
state matrix A by the similarity conversion

(20.11) $A = S\Lambda S^{-1}.$

Note that from conditions (20.3) and rank $(Z_k) = 1$, the above statement is equivalent to finding the <u>same</u> nonzero column in <u>each</u> of the component matrices of A.

Also observe that through the output function $y = Cx$, it is constant multiples of entries of the Z_k that are directly being measured since the solution x is expressed in component form $(x(0) = [0, \ldots, 0, 1, 0, \ldots 0]^T$ where 1 is in the jth position)

$$(20.12) \qquad x_i(t) = \sum_{k=1}^{n} z_{ij}^{(k)} \exp(\lambda_k t)$$

for some j (see [143] for the application of component matrices to the solution of systems of linear differential equations).

Another strategy sufficient for unique identification of A is obtained through interchanging the words rows and columns in the preceding condition associated with (20.11). One is then dealing with left eigenvectors of A. If J is the (row) modal matrix of linearly independent left eigenvectors formed analogously to S, then A is determined by

$$(20.13) \qquad A = J^{-1} \Lambda J.$$

The question now arises: Just what equations are available to relate the unknown entries in A and Z_k to the known entries in Z_k, $k = 1, 2, \ldots, n$? We already have a number of conditions above, such as (20.2), (20.4), (20.6), and (20.9). Extra equations arise from the use of the linearly independent functions $f(\lambda) = \lambda, \lambda^2, \lambda^3, \ldots,$ in (20.1) together with prior knowledge of certain entries $a_{ij}^{(k)}$ of A^k, particularly that some are zero from the specific description of the model being analyzed. For instance, if there is not a path in the connectivity diagram of the system by which compartment i can be reached from compartment j in k compartment transitions, then $a_{ij}^{(k)} = 0$ and the identity

$$(20.14) \qquad A^k = \lambda_1^k Z_1 + \lambda_2^k Z_2 + \ldots + \lambda_n^k Z_n$$

yields the single scalar equation

$$(20.15) \qquad 0 = a_{ij}^{(k)} = \lambda_1^k z_{ij}^{(1)} + \ldots + \lambda_n^k z_{ij}^{(n)} \; .$$

This linear equation in $z_{ij}^{(1)}, \ldots, z_{ij}^{(n)}$ vividly demonstrates an advantage of the component matrix method -- nonlinearities in the a_{ij} that occur in powers of A now show up in (20.15) as nonlinearities only in the λ_i, which are considered to be observables. Furthermore, additional equations come about from knowledge of compartmental excretion. If a_{0j} is known, in particular if compartment j has no excretion, then summation of the jth column of (20.10) gives

$$(20.16) \qquad - a_{0j} = \sum_{i=1}^{n} \left(\lambda_1 z_{ij}^{(1)} + \ldots + \lambda_n z_{ij}^{(n)} \right),$$

yet another linear equation in the $z_{ij}^{(k)}$.

It should be noted that introducing the component matrices increases the number of unknowns in the problem, however there is a tradeoff in that all of the above equations for the $z_{ij}^{(k)}$ are linear with the single exception of the quadratic system equations (20.2). This simplifies the form of the equations to the extent where the effects of any redundancy in the equations and a priori knowledge of certain rate constants can be more readily analyzed.

20D. Identification of a Lipoprotein Model

An example will serve to crystallize these thoughts [37]. Consider a two compartment model in which compartment 1 is perturbed by a unit impulse input, its response is observed, and the excretion rate constant a_{02} is known to be zero. Let us suppose that the measured response of the system via compartment 1 is

$$(20.17) \qquad y(t) = x_1(t) \approx 0.689 \exp(- 2.823t) + 0.311 \exp(- 0.177t).$$

Since the observed eigenvalues $\lambda_1 \equiv - 2.823$ and $\lambda_2 \equiv - 0.177$ of the compartmental matrix A are distinct, the fundamental matrix of the system $\dot{x} = Ax$ can be written

$$(20.18) \qquad \Phi(t) \equiv \exp(tA) = \exp(\lambda_1 t)Z_1 + \exp(\lambda_2 t)Z_2$$

in keeping with representation (20.1). Since $x_0 \equiv [1 \ \ 0]^T$, then the solution $x = \Phi x_0$ of the system yields

$$(20.19) \qquad x_1(t) = z_{11}^{(1)} \exp(\lambda_1 t) + z_{11}^{(2)} \exp(\lambda_2 t).$$

Comparing (20.17) and (20.19), the observed response indicates that $z_{11}^{(1)} \equiv 0.689$

and $z_{11}^{(2)} \equiv 0.311$ are known. We now wish to determine either a complete row or column in each of Z_1 and Z_2. This can be accomplished through the following steps. The equation $Z_1 Z_1 = Z_1$ of (20.2) and its (1, 2)-entries give

$$z_{11}^{(1)} z_{12}^{(1)} + z_{12}^{(1)} z_{22}^{(1)} = z_{12}^{(1)} .$$

If we assume that $z_{12}^{(1)}$ is nonzero, then $z_{22}^{(1)} = 1 - z_{11}^{(1)} = 0.311$. Condition (20.9) implies the relationship

$$1 = z_{22}^{(1)} + z_{22}^{(2)}$$

from which there results $z_{22}^{(2)} = 0.689$. Note that expression (20.15) is not useful to us in this example. However, (20.16) when $j = 2$ yields

$$0 = \left[\lambda_1 z_{12}^{(1)} + \lambda_2 z_{12}^{(2)} \right] + \left[\lambda_1 z_{22}^{(1)} + \lambda_2 z_{22}^{(2)} \right] .$$

The quantity within the second pair of brackets is presently known; what is needed is another equation in $z_{12}^{(1)}$ and $z_{12}^{(2)}$. Such an equation presents itself in (20.9):

$$z_{12}^{(1)} + z_{12}^{(2)} = 0 .$$

Simultaneous solution of the two preceding linear equations yields $z_{12}^{(1)} = -0.378$ and $z_{12}^{(2)} = 0.378$. With these values we have now determined row one in each of Z_1 and Z_2. Hence the compartmental matrix A is uniquely identifiable. For, if we let

$$J \equiv \begin{bmatrix} z_{11}^{(1)} & z_{12}^{(1)} \\ \\ z_{11}^{(2)} & z_{12}^{(2)} \end{bmatrix}$$

then

$$A = J^{-1} \Lambda J = \begin{bmatrix} -2. & 1. \\ \\ 1.5 & -1. \end{bmatrix} .$$

Perhaps a quicker route to the identification of A in this example is through noting that the model is a SEC system and that the compartment being measured is also the only one in the system with an exit. Hence Hamilton's formula (16.23) applies to give

$$a_{01} = - \left[z_{11}^{(1)}/\lambda_1 + z_{11}^{(2)}/\lambda_2 \right]^{-1} = 0.5.$$

Once a_{01} is known, then condition (20.16) gives

$$- 0.5 = \left[\lambda_1 z_{11}^{(1)} + \lambda_2 z_{11}^{(2)} \right] + \left[\lambda_1 z_{21}^{(1)} + \lambda_2 z_{21}^{(2)} \right].$$

A second equation for $z_{21}^{(1)}$ and $z_{21}^{(2)}$ is

$$0 = z_{21}^{(1)} + z_{21}^{(2)}$$

from (20.9). Solution of the preceding two equations results in $z_{21}^{(1)} = - 0.567$ and $z_{21}^{(2)} = 0.567$. At this stage, the first column in each of Z_1 and Z_2 are known and thus A is uniquely identifiable from the (column) modal matrix

$$S \equiv \begin{bmatrix} z_{11}^{(1)} & z_{11}^{(2)} \\ \\ z_{21}^{(1)} & z_{21}^{(2)} \end{bmatrix}.$$

The final calculation is $A = SAS^{-1}$.

Exercise 20.7.   In the above example, show that all entries in Z_1 and Z_2 can be determined. Note that this provides an alternate (but longer) identification method for A via the formula $A = \lambda_1 Z_1 + \lambda_2 Z_2$.

20E.   The Identification Technique Using Modal Matrices

For the identification of higher-ordered problems it may be advantageous to shift over to using the quantities s_{ij} and $s_{ij}^{(-1)}$ instead of the entries $z_{ij}^{(k)}$ of the component matrices. Through $y = Cx$, certain $z_{ij}^{(k)}$ are observed directly. These can be converted into statements involving entries in the matrices S and S^{-1} by formula

(20.5). Also via the relationship $A^k = S\Lambda^k S^{-1}$, condition (20.15) is replaced by

(20.20) $a_{ij}^{(k)} = <r_i, \Lambda^k c_j>$ $k = 1, 2, \ldots$,

where $<\cdot, \cdot>$ denotes the usual inner product of vectors, and r_i and c_j are row i of S and column j of S^{-1}, respectively. Likewise condition (20.16) passes over to

(20.21) $- a_{0j} = \sum_{i=1}^{n} a_{ij} = \sum_{i=1}^{n} <r_i, \Lambda c_j>$ $j = 1, 2, \ldots, n$

where we are also using (20.20) when $k = 1$. Moreover, the last two equations are accompanied by the condition

(20.22) $<r_i, c_j> = \delta_{ij}$ $i, j = 1, 2, \ldots, n.$

20F. Identification of a Pharmacokinetic System

The following three-compartment experiment [172, p. 107] illustrates the usage of the quantities s_{ij} and $s_{ij}^{(-1)}$ in the identification of a model. A pharmacokinetic system has connectivity diagram as in Figure 20.1. In the linear system (16.5), let us try the experiment where

$$B \equiv [1 \quad 0 \quad 0]^T \qquad\qquad C \equiv [0 \quad 0 \quad 1].$$

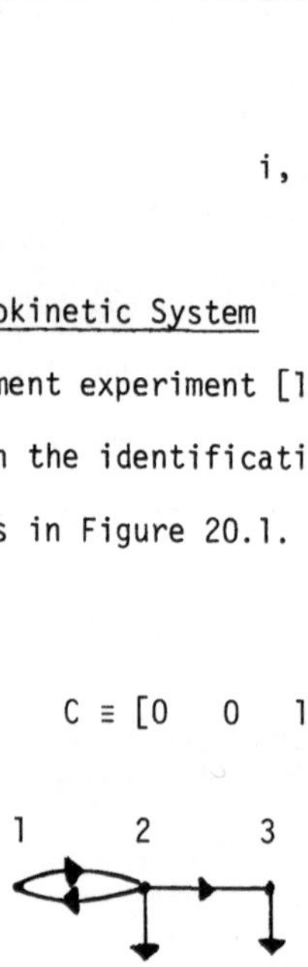

Figure 20.1.

Is this experiment uniquely identifiable? Let us check the nonidentifiability condition of Section 18 first. From Figure 20.1 it is clear that there are $\mu = 5$ unknown rate constants. Direct computation of the number of equations available from $\hat{\phi}(s)$ to determine these unknowns is

$$N = (n-m) + n(3,1) - p(3,1) = 3 - 0 + 3 - 2 = 4.$$

Therefore since the nonidentifiability condition $N < \mu$ is met, we move on to the more complex experiment in which

$$B \equiv [1 \quad 0 \quad 0]^T \qquad C \equiv \begin{bmatrix} 1 & 0 & 0 \\ 0 & 0 & 1 \end{bmatrix}.$$

In this case, model (16.5) represents a feasible experiment because $N = 6$. In fact, we get unique identification of A as follows. The output function $y(t)$ is $[x_1(t) \quad x_2(t)]^T$ or

$$\left[\sum_{k=1}^{3} z_{11}^{(k)} \exp(\lambda_k t) , \qquad \sum_{k=1}^{3} z_{31}^{(k)} \exp(\lambda_k t) \right]^T.$$

We take as knowns λ_k, $z_{11}^{(k)}$, and $z_{31}^{(k)}$, $k = 1, 2, 3$. Thus if each $z_{21}^{(k)}$ can be determined, we are finished. This is accomplished through identifying the second row, r_2, of S. Because the scaling of each column of S is arbitrary, the first row of S can be taken for convenience as $r_1 \equiv [1 \quad 1 \quad 1]$ provided all modes appear in the measured state variable. Then c_1 follows as the column vector

$$c_1 = \left[s_{11}^{(-1)} \quad s_{21}^{(-1)} \quad s_{31}^{(-1)} \right]^T = \left[z_{11}^{(1)} \quad z_{11}^{(2)} \quad z_{11}^{(3)} \right]^T$$

of coefficients of the exponentials in x_1 by formula (20.5). Hence the s_{3i}-values, or the third row entries of S, are derived from the equations

$$s_{3i} \, s_{i1}^{(-1)} = z_{31}^{(-i)} \qquad\qquad i = 1, 2, 3.$$

Utilizing equations (20.20) - (20.22), identification of the model by successive linear stages yields column vector c_3 from the system

$$< r_1, c_3 > \ = 0$$

$$< r_1, \Lambda c_3 > \ = a_{13} = 0$$

$$< r_3, c_3 > \ = 1$$

and then r_2 from the set of equations

$$< r_2, c_1 > = 0$$

$$< r_2, c_3 > = 0$$

$$\sum_{i=1}^{3} < r_i, \Lambda c_1 > = - a_{01} = 0.$$

Now S is completely known and so A is uniquely determined from $A = S\Lambda S^{-1}$.

Exercise 20.8.   Compare the analysis for the above example with our previous identification technique using the transfer function matrix. Discuss redundancy of any transfer function coefficient equations.

Exercise 20.9.   By the component matrix method, attempt to identify the compartmental system associated with Figure 16.3 wherein there is a unit bolus input to the second compartment and then the third compartment is sampled over time.

20G. Spectral Sensitivity of a Linear Model

    Let us return for a moment to further examine the implications of the interesting relationship (20.5) between the entries of the component matrices and the column modal matrix and its inverse. As the next theorem indicates, this formula allows a very nice interpretation of the individual elements $z_{ji}^{(k)}$ of the component matrices.

THEOREM 20.1.   Let i, j be arbitrary but fixed and suppose E_{ij} is the $n \times n$ matrix with unity in its (i,j)- position and zeros elsewhere. For the $n \times n$ matrix A with distinct eigenvalues $\{\lambda_k\}$, define

$$A(\varepsilon) \equiv A + \varepsilon E_{ij}, \qquad 0 < |\varepsilon| \ll 1.$$

Then the kth eigenvalue $\lambda_k(\varepsilon)$ of $A(\varepsilon)$ possesses the regular perturbation expansion

$$(20.23) \qquad \lambda_k(\varepsilon) = \lambda_k + \lambda_k^{(1)} \varepsilon + \lambda_k^{(2)} \varepsilon^2 + \ldots,$$

where the coefficient $\lambda_k^{(1)} = z_{ji}^{(k)}$.

Proof.   The existence of the power series expansion in ε for nonrepeated eigenvalues is shown by Lancaster [143, p. 241]. Then, according to [143, p. 245], the first order perturbation coefficient has the formula

$$(20.24) \qquad \lambda_k^{(1)} = \,,$$

where f_k is a left eigenvector of A corresponding to λ_k, and e_k is a right eigen-vector which is the kth column of the column modal matrix S. Due to orthonormality, the vector f_k^T is the same as the kth row of S^{-1}. Hence the inner product in (20.24) yields

$$\lambda_k^{(1)} = s_{jk}s_{ki}^{(-1)} .$$

But then relationship (20.5) implies that $\lambda_k^{(1)} = z_{ji}^{(k)}$.   ///

If (20.23) is rewritten as

$$[\lambda_k(\varepsilon) - \lambda_k]/\varepsilon = \lambda_k^{(1)} + O(\varepsilon),$$

then Theorem 20.1 tells us that the entries in the component matrices have the interpretation

$$(20.25) \qquad \partial\lambda_k/\partial a_{ij} = z_{ji}^{(k)} \qquad\qquad i, j, k = 1, 2, \ldots, n,$$

that is, they give a measure of the sensitivity of the kth eigenvalue to changes in the model parameter a_{ij} alone.

Exercise 20.10.   Demonstrate the above theory for the compartmental matrix of Exercise 20.1 in the case of a small perturbation of $\varepsilon \equiv -0.1$ in a_{21}, and how it affects the eigenvalue λ_2. Check equation (20.25) by actually computing $\lambda_2(\varepsilon)$ and then forming $[\lambda_2(\varepsilon) - \lambda_2]/\varepsilon$.

A different route to (20.25) is taken by Caswell [55] in a discussion of the problem of sensitivity of population growth rate to small changes in the Leslie matrix model parameters. There the model predicts the population will grow geo-metrically at a rate λ, where λ is the positive real eigenvalue of the Leslie matrix A, of modulus greater than any other (see Anderson [3]). If the corresponding left and right eigenvectors are $f \equiv [u_i]$ and $e \equiv [v_j]$, then Caswell arrives at the con-clusion (his equation (10))

(20.26) $d\lambda/da_{ij} = u_i v_j$.

__Exercise 20.11.__   Verify that Caswell's formula in (20.26) is in fact the same as that given in (20.25).

We close this section by noting that properties of component matrices of a compartmental matrix have not been fully explored. It is an open area which merits additional research. With regard to identification, it would appear that the _nonlinear_ Markov parameter equations of Theorem 17.3 may be replaced by the _linear_ equations

$$\phi^{(k)}(0) = \lambda_1^k \, CZ_1 B + \ldots + \lambda_n^k \, CZ_n B,$$

k = 1, 2, ..., n-1, via (20.14), but we must still live with certain quadratic equations (those in (20.2)) which correspond in some way to the remaining nonlinear equations of (17.5).

__Exercise 20.12.__   If A is a diagonalizable compartmental matrix, show that

$$\lambda_k = tr(Z_k A) = tr(A Z_k)$$

for all k.

__Exercise 20.13.__   Let A be a diagonalizable compartmental matrix with no eigenvalue equal to 0 or - 1. Prove that

$$(A + Z_k)^{-1} = A^{-1} - Z_k/\lambda_k(\lambda_k + 1).$$

When is $A + Z_k$ invertible?

__Exercise 20.14.__   If A is an $n \times n$ nonsingular diagonalizable compartmental matrix, prove that

$$Z_n = \lim_{k \to \infty} (\lambda_n A^{-1})^k \geq 0 .$$

__Exercise 20.15.__   Suppose A as in Exercise 20.14 has $\lambda_n < - 1$. Then show that $(A + Z_n)^{-1} \leq 0.$

SECTION 21. CONTROLLABILITY, OBSERVABILITY, AND PARAMETER IDENTIFIABILITY

21A. The Control Problem

Given the compartmental system (A, B, C), we would like to know the conditions

under which control can be exercised over the chosen states. The input function u

is prescribed and the structure in B determines to what extent the system behavior

can be modified. The output function y is measured, and the structure in C indi-

cates the type of information available for observation. The objective in the

control problem is to design systems which will achieve the desired output in

response to known inputs. These input-output components of the model (A, B, C)

interface through the system connections which are represented as nonzero a_{ij} in

the compartmental matrix A.

21B. Completely Controllable Systems

Two modern qualitative concepts in control theory, discussed by Kalman [137],

[138] and which characterize the implications of input-output structure, are con-

trollability and observability. Basically speaking, controllability investigates

the possibility of guiding the state variable from control of the input. On the

other hand, observability studies our ability to estimate the state from the output

information.

The time-invariant system $\dot{x}$ = Ax + Bu is said to be completely controllable

(CC) if for the initial state vector x_0 and any prescribed terminal state vector

x_1, there is a positive time t_1 and an input u(t) applied over the time interval

[0, t_1] such that $x(t_1)$ = x_1. Intuitively then, the system is CC if we are able

to "control" the system state, i.e., if an input u can be constructed which will

drive the state variable to any preselected state in a finite amount of time. The

terminology "completely" refers to the fact that all states can be reached by input.

To get a better understanding of complete controllability and in fact to get

a more easily applied criterion for CC in terms of the information in A and B (we

have not included the output matrix C because it is irrelevant for the study of

controllability), a direct appeal is made to the solution (10.5) of the differential equation model subject to the initial condition $x(0) = 0$ (the choice of $x_0 \equiv 0$ is one of convenience):

(21.1) $\qquad x(t) = \int_0^t \exp[(t-s)A]\ Bu(s)ds.$

Now, for prescribed x_1, we need to produce a time $t_1 > 0$ and an input $u(t)$ on $[0,t_1]$ such that the solution function of (21.1) satisfies $x(t_1) = x_1$. This can be accomplished as follows. Select an arbitrary but fixed $t_1 > 0$ and set the $q \times 1$ input vector to be

(21.2) $\qquad u(t) \equiv B^T \exp(-A^T t)K^{-1} \exp(-At_1)x_1$

where the $n \times n$ matrix K is defined by

$$K \equiv \int_0^{t_1} \exp(-As)\ BB^T \exp(-A^T s)ds.$$

Then by substituting expression (21.2) into (21.1),

$$x(t_1) = \exp(t_1 A)\ [\int_0^{t_1} \exp(-sA)\ BB^T \exp(-A^T s)ds]K^{-1} \exp(-At_1)x_1$$

$$= \exp(t_1 A)KK^{-1} \exp(-At_1)x_1$$

$$= x_1.$$

Thus input (21.2) steers the state vector from zero initially to x_1 at time t_1.

The single condition on which the above construction depends is that K be a nonsingular matrix. To prove this invertibility of K, suppose there is an $n \times 1$ vector z such that $Kz = 0$. It is then sufficient to show that $z = 0$. Now

(21.3) $\qquad 0 = z^T Kz = \int_0^{t_1} c^T(s)\ c(s)ds$

in which $c(t)$ is the $q \times 1$ vector $B^T \exp(-A^T t)z$. The vector inner product $c^T c$ in (21.3) must be nonnegative. Hence for condition (21.3) to hold, the integrand must vanish identically on $[0,t_1]$. In particular, when t is set equal to zero, $z^T B = 0$. Successive derivatives $d^k c^T(t)/dt^k$ evaluated at $t = 0$ yield

$$z^T A^k B = 0, \qquad k = 1, 2, \ldots, n-1.$$

Thus z is orthogonal to all columns of the partitioned n × (nq) matrix

(21.4) $\qquad U \equiv [B, AB, A^2 B, \ldots, A^{n-1} B].$

Now rank (U) ≤ n. It must follow that z = 0 provided U has full rank n.

The matrix U defined in (21.4) is called the <u>controllability matrix</u> of the pair of matrices (A, B). Observe that if there is a single input to the system (q = 1), then U is a square n × n matrix so that full rank is equivalent to invertibility of U. We have just motivated the entry of this matrix into the controllability problem by proving one direction of the equivalence given as the next theorem [137], [57, p. 177], [138], [153, p. 282].

<u>THEOREM 21.1.</u>   The system $\dot{x}$ = Ax + Bu is CC if and only if the associated controllability matrix (21.4) is of full rank n.

As an illustration, consider the compartmental model

(21.5) $\qquad \dot{x} = \begin{bmatrix} a_{11} & a_{12} \\ a_{21} & a_{22} \end{bmatrix} x + \begin{bmatrix} 1 \\ 0 \end{bmatrix} u, \qquad t \geq 0,$

with $a_{11} \equiv - a_{21}$ and $a_{22} \equiv - a_{12} - a_{02}$. Here

$$B = [1 \quad 0]^T \qquad\qquad AB = [a_{11} \quad a_{21}]^T$$

so that the controllability matrix is

$$U = [B, AB] = \begin{bmatrix} 1 & a_{11} \\ 0 & a_{21} \end{bmatrix}$$

and is of rank 2 since $\det U = a_{21} \neq 0$. Hence, according to Theorem 21.1, the system (21.5) is completely controllable.

Exercise 21.1.   Suppose we have a CC single input ($q = 1$) compartmental system (A, B, C) wherein every compartment is individually sampled (so take $C \equiv I$). Show that A and B are identifiable from knowledge of the Markov parameters.

21C. Completely Observable Systems

The dynamical system $\dot{x} = Ax + Bu$, $y = Cx$ (we really don't need the input distribution matrix B for this definition but leave it in for generality) is said to be completely observable (CO) if there exists a positive time t_1 such that knowledge of the output function $y(t)$ for all t in $[0,t_1]$ suffices to determine the initial state x_0 of the system. Here we suppose that the matrices A and C are known and the system outputs are known, but the initial state $x(0)$ is unknown until the outputs become available. From measuring the outputs, the initial value $x(0)$ can be inferred provided the system is CO.

The criterion for CO in terms of the matrices of the system is analogous to that of CC.

THEOREM 21.2.   The system $\dot{x} = Ax + Bu$, $y = Cx$ is CO if and only if its associated $(pn) \times n$ observability matrix

$$(21.6) \qquad V \equiv \begin{bmatrix} C \\ CA \\ CA^2 \\ \cdot \\ \cdot \\ \cdot \\ CA^{n-1} \end{bmatrix}$$

has full rank n.

Let us recall the example of (21.5) to which is added $C \equiv [1 \quad 1]$. Then V is square,

$$(21.7) \qquad V = \begin{bmatrix} 1 & 1 \\ & \\ 0 & -a_{02} \end{bmatrix},$$

and has determinant value - a_{02}. Thus this pair of matrices (A, C) is CO by Theorem 21.2.

The concepts of CC and CO should be considered from a practical as well as a mathematical perspective. It is unusual that for a specific compartmental model the controllability or observability of the model is not clear. Often our intuitive knowledge of the system precludes the need to laboriously proceed through all of the formal calculations involved in theorems like those above. If the system is CC, then we will see all modes of the differential equations being excited from the input. On the other hand, if the system is CO, then all modes of the equations can be observed at the output.

21D. Realizations and Identifiability

At this point we would like to discuss how the concepts of controllability and observability relate to the realization and identification problems for (A, B, C).

Let us say that the system S_1, meaning (A_1, B_1, C_1), and the system S_2, that is, (A_2, B_2, C_2), are _equivalent_ if for any initial state x_0^1 of S_1 there is at least one initial state x_0^2 of S_2 such that the output functions $y(t; S_1) \equiv y(t; S_2)$ for all $t > 0$ and inputs u [227], [228]. As we will see, one way of getting a system which is equivalent to an existing system (A_1, B_1, C_1) is through a linear change of variable (this technique was used in Section 9 for tracer concentration and specific activity equations). Indeed, suppose P is a $n \times n$ constant nonsingular matrix and z(t) is a $n \times 1$ vector function. We set $x(t) \equiv Pz(t)$ in the system

$$\dot{x} = A_1 x + B_1 u$$

(21.8) $$x(0) = x_0^1$$

$$y = C_1 x.$$

This transforms (21.8) into the form

$$\dot{z} = (P^{-1}A_1 P)z + (P^{-1}B_1)u$$

$$z(0) = P^{-1}x_0^1$$

$$y = (C_1 P)z .$$

(Note that if $x_0^1 = 0$, then $z(0) \equiv z_0 = 0$.) Define

(21.9) $A_2 \equiv P^{-1}A_1P$, $B_2 \equiv P^{-1}B_1$, $C_2 \equiv C_1P$.

Any two equivalent systems which are related as in (21.9) are called <u>algebraically</u> <u>equivalent systems</u>.

<u>Exercise 21.2.</u>   Prove that CC and CO of the linear dynamic system (A, B, C) are invariant under any equivalence transformation.

It now follows that if S_1 is algebraically equivalent to S_2 under zero initial conditions, then the systems do have the same output. For, from the solution of the model,

$$y(t; S_2) = C_2 x(t; S_2)$$

$$= \int_0^t C_2 \exp[(t-s)A_2] B_2 u(s)ds$$

(21.10) $$= \int_0^t C_1 P \exp[(t-s)P^{-1}A_1P] P^{-1} B_1 u(s)ds$$

$$= \int_0^t C_1 \exp[(t-s)A_1] B_1 u(s)ds$$

$$= C_1 x(t; S_1) = y(t; S_1).$$

<u>Exercise 21.3.</u>   Verify the equality of (21.10).

The concept of algebraic equivalence has also been previously demonstrated in the <u>realization problem</u> (Exercise 16.3): for a given impulse response function ϕ, find a triple of matrices (A, B, C) which generate ϕ in that $\phi(t) = C\exp(tA)B$. We recall that such a triple is a <u>realization of</u> ϕ. The dimension of a realization (A, B, C) of ϕ is the dimension of A, that is, the number of compartments in the compartmental system. There is a <u>smallest</u> dimension $M(\phi)$ [44] attained by realizations of this prescribed ϕ, and realizations of order $M(\phi)$ are said to be <u>irreducible</u>. For instance, if $\phi(t) = 1$ for all $t \geq 0$, then the system (A_1, B_1, C_1) given by

$$\begin{bmatrix} -1 & 0 \\ 0 & 0 \end{bmatrix}, \quad \begin{bmatrix} 0 \\ 1 \end{bmatrix}, \quad [0 \quad 1],$$

has ϕ as its impulse response function. However, this triple is <u>not</u> an irreducible realization becuase the 1×1 matrices

$$(A_2, B_2, C_2) \equiv ([0], [1], [1])$$

also realize $\phi(t) = 1$. Thus $M(\phi) = 1$.

Kalman [137], [57, p. 203] noted that a realization is irreducible if and only if it is both CC and CO. As an example, the compartmental system of (21.5) and (21.7) is both CC and CO, and hence it follows that the system is irreducible. Kalman [137], [44], [57] also pointed out that all irreducible realizations are algebraically equivalent.

A number of misconceptions have arisen about realizations which are irreducible (CC and CO). For example, it has been stated that strongly connected compartmental systems are always CC and CO. However, this is just not the case. The strongly connected 2-compartment system whose dynamic description is

$$(21.11) \quad \dot{x} = \begin{bmatrix} -2. & 0.5 \\ 1. & -2.5 \end{bmatrix} x + \begin{bmatrix} 0.5 \\ 0.5 \end{bmatrix} u$$

is not CC since $\det[B, AB] = 0$.

<u>Exercise 21.4.</u> Given a compartmental matrix A, suppose the pair (A, B) is a CC pair of matrices for <u>any</u> choice of B. Prove that the compartmental system must be strongly connected. Also show that if the pair (A, C) is a CO pair of matrices for <u>every</u> C, then the compartmental system is strongly connected.

A compartmental system can be non CC and/or non CO either because of its own overall structure or due to a particular combination of parameters, as in the example of (21.11). In the former case, the matrix U and/or V always has less than full rank for whatever values of the model parameters relative to a system of fixed structure. It means there are no paths between a certain input and a certain state or between a certain state and a certain output.

Also in the past, it has been thought by some scientists that irreducibility of (A, B, C) is equivalent to the structural identifiability of that system. But CC and CO are neither necessary nor sufficient conditions for identifiability [60]. Let us scrutinize this more closely. We first observe that CC and CO are not sufficient to guarantee identifiability of the system [82, p. 122]. The model given in (21.5) and (21.7) is CC and CO as we have already established. But if the non-identifiability test affiliated with Theorem 18.5 is applied, then the number of independent equations available from the transfer function is $\nu \leq N - 1 = 2 \equiv \beta$. The number of unknown model parameters is $\mu = 3$. Hence $\beta < \mu$ confirms that the model is nonidentifiable in spite of the fact that the system is both CC and CO.

On the other hand, system identifiability does not imply CC and/or CO of the system. For, let us investigate the controllability, observability, and identifiability of the system (A, B, C) where

$$A \equiv \begin{bmatrix} a_{11} & 0 & 0 \\ a_{21} & 0 & 0 \\ a_{31} & 0 & 0 \end{bmatrix} \qquad a_{11} \equiv - a_{21} - a_{31}$$

$$B^T \equiv [1 \quad 0 \quad 0] \qquad C \equiv [0 \quad c_1 \quad c_2]$$

and in which c_1, c_2 are unequal known constants. The transfer function $\hat{\phi}$ is a scalar since $p = q = 1$. From the denominator of $\hat{\phi}$, just one equation $s_1 = a_{21} + a_{31}$ is available for identification purposes since there are two sinks in the compartmental system. The only other parameter equation comes from the numerator of $\hat{\phi}$,

$$\Gamma_1 = M_1 + s_1 M_0 = c_1 a_{21} + c_2 a_{31} \; ,$$

which is independent of the previous equation since $c_1 \neq c_2$. Therefore the model is structurally identifiable because c_1 and c_2 are different. However, upon forming

the controllability matrix U and the observability matrix V, it is found that both
are of rank two, and so the system (A, B, C) is neither CC nor CO.

Exercise 21.5. [89].   Discuss the controllability, observability, and identifiability of the system composed of

$$A \equiv \begin{bmatrix} a_{11} & 0 & 0 \\ 0 & a_{22} & 0 \\ a_{31} & a_{32} & 0 \end{bmatrix} \qquad \begin{array}{l} a_{11} \equiv - a_{31} \\ \\ a_{22} \equiv - a_{32} \end{array}$$

$$B^T \equiv C \equiv \begin{bmatrix} 1 & 0 & 0 \\ 0 & 1 & 0 \end{bmatrix} .$$

Of course, there are situations in which CC and/or CO of the compartmental
system are sufficient to give structural identifiability. Common to tracer experiments is the single-input case. We will also assume in the hypothesis of the next
theorem the sampling of enough compartments in the system so as to provide knowledge of the controllability matrix. There certainly are instances where this
occurs. In the 4-compartment liver disease model of Figure 16.4, the first, second,
and fourth compartments are sampled and single input is to the first compartment.
The controllability matrix is then

$$U = \begin{bmatrix} 1 & a_{11} & a_{11}^{(2)} & a_{11}^{(3)} \\ 0 & a_{21} & a_{21}^{(2)} & a_{21}^{(3)} \\ 0 & a_{31} & a_{31}^{(2)} & a_{31}^{(3)} \\ 0 & 0 & a_{41}^{(2)} & a_{41}^{(3)} \end{bmatrix} .$$

Close examination of the equations after Figure 16.4 show that all entries of U are

known from either the Markov parameters M_1, M_2, M_3, or the outflow equations
$a_{11} = - a_{21} - a_{31}$ and $a_{33} = - a_{13} - a_{43}$.

THEOREM 21.3.   A single-input CC compartmental system is structurally identifiable
if its controllability matrix is known.

Proof.   The verification of this result discloses an interesting matrix structure
following an equivalence transformation involving the controllability matrix U.
The matrix B is part of U, so that B is assumed to be known.   Since C does not enter
into CC, we will suppose that C is given so that it remains only to determine A in
the system (A, B, C).

Since the system has a single input, the input distribution matrix B is $n \times 1$
thus providing a $n \times n$ matrix $U = [B, AB, A^2B, \ldots, A^{n-1}B]$.   Complete controllability
of the system is then equivalent to invertibility of U and so this matrix can be
used to affect a change of variable in (A, B, C) of (21.8) by the transformation
$x = Uz$.   The transformed system $(U^{-1}AU, U^{-1}B, CU)$ is CC by Exercise 21.2 and possesses an especially simple structure for the first two of its matrices [153,
p. 240].   The matrix

$$U^{-1}B = [1 \quad 0 \quad \ldots \quad 0]^T ,$$

and the matrix $\overline{A} \equiv U^{-1}AU$ has the form of a __companion matrix__, with ones on the sub-
diagonal and a last column made up of the coefficients in the characteristic
equation of A:

$$- [s_n \quad s_{n-1} \quad \ldots \quad s_1]^T .$$

Thus $\overline{A}$ consists only of known constants and observables from knowledge of the eigenvalues of the system.   Therefore since U is given, the matrix A is uniquely determined by the similarity computation $A = U\overline{A}U^{-1}$, and the proof is complete.   (Note
that all eigenvalue information of the system is wrapped up in $\overline{A}$, whereas U contains
the entirety of the Markov parameter information.)

///

A particular case of this theorem occurs when <u>all</u> of the compartments in the system are independently observed. Without loss of generality we can take $C \equiv I$ (note then that CO of the system is implicitly assumed). Then from knowledge of the impulse response ϕ, we have the Markov parameters

$$\phi(0) = CB = B$$

$$\phi^{(1)}(0) = CAB = AB$$

$$\vdots$$

$$\phi^{(n-1)}(0) = CA^{n-1}B = A^{n-1}B \ ,$$

and hence know the entries of the controllability matrix U. This special case of Theorem 21.3 is presented by Bellman and Astrom [24], although they fail to include the necessary single-input requirement on the system. Their proof is somewhat different in that it does not use an equivalence transformation.

A slight variation of this $C = I$ case is present in the liver model considered just above. Since the system of Figure 16.4 is closed to the outside environment (i.e., $a_{0j} = 0$ for all j), tracer mass is conserved. Thus we get a measure of the third compartment (liver) by the formula

$$x_3(t) = D - x_1(t) - x_2(t) - x_4(t)$$

at any time t which is a common sampling time for each of compartments 1, 2, and 4. Thus for all practical purposes, we can envision measurement of all compartments and thus take $C = I$. Hence the controllability matrix

$$U = [B, AB, A^2B, A^3B] = [M_0, M_1, M_2, M_3]$$

is determined from knowledge of the first $n - 1$ Markov parameters.

Exactly parallel procedures can be worked out for a CO single-output system where the change of variable is based on the nonsingularity of the observability matrix [153, p. 294]. This yields the following theorem, which is the "dual" of

Theorem 21.3 (again, a special case of this result is when B = I [93, p. 454]).
Also see [78].

THEOREM 21.4.   A single-output CO compartmental system is structurally identifiable
if its observability matrix is known.

21E.   A Third Method of Identifiability

Let us now suppose we have at hand a compartmental system with zero initial
condition which is both CC and CO [244].   What does it take in addition for the
system to be structurally identifiable?   One approach towards answering this question is to note that since the compartmental system is irreducible, then any one
system with which we work is algebraically equivalent to another [137].   In the
equivalence transformation from compartmental system (A_1, B_1, C_1) of (21.8) to form
(21.9), we would desire the coefficient matrix A_2 to be compartmental with the same
structure as A_1, and the input, output matrices B_2, C_2 to be nonnegative so that the
new system (A_2, B_2, C_2) is admissible from the compartmental analysis viewpoint.
Thus a natural question that arises is what conditions must fall on the invertible
$n \times n$ matrix P so that the new system is in fact an admissible compartmental model?
Certainly one matrix that falls within this class of invertible matrices is P = I.

For the identification process we must initially eliminate those systems that
do not satisfy the a priori conditions imposed on the model, such as the number of
compartments, which fractional transfer coefficients are zero, what compartments
have excretion to the outside environment, etc.   These conditions establish classes
of matrices H_A, H_B, H_C to which each of the matrices A, B, C, respectively, of a
system (A, B, C) must belong [227], [228].   For instance, if an irreducible realization (A, B, C) exists which is of dimension two, for which $a_{12} = 0$ and compartment
1 has no excretion, then class H_A may consist of all 2×2 compartmental matrices
$A \equiv [a_{ij}]$ with $a_{12} = 0$, $a_{11} = -a_{21} < 0$, $a_{22} < 0$, but otherwise arbitrary values.
By defining the sets H_A, H_B, and H_C, all internal couplings of the system are being
specified.

Exercise 21.6. Consider the above example where

$$H_A \equiv \left\{ A = \begin{bmatrix} -a & 0 \\ a & -b \end{bmatrix} : a,b > 0 \right\}$$

that is, H_A is the collection of all 2 × 2 compartmental matrices compatible with
the connectivity diagram of Figure 17.2. Discuss restrictions that must apply to
an invertible 2 × 2 matrix P so that $A \in H_A$ implies $P^{-1}AP \in H_A$. Is it possible to
have a nondiagonal transformation matrix P?

For the irreducible compartmental system (A, B, C) with zero initial condition
to be structurally identifiable, there must be only a single candidate matrix in
each of the classes H_A, H_B, and H_C; i.e., P is forced to be the identity matrix.
To illustrate, let us consider an example which is discussed by Cobelli et al. [66,
p. 149] and Travis et al. [227], [228]. Here we investigate the role of the kidney
in maintaining the optimal chemical composition of bodily fluids. One process which
involves the regulation of the internal environment by the kidney is the filtration
of the blood plasma by the glomeruli. In this input-output experiment, tracer is
injected into the renal artery. It is assumed that the tracer molecule is large
enough so as not to be filtered and passed in the urine. Thus the amount of tracer
in the urine output of Figure 21.1 is negligible. We record the single-pass disappearance of tracer from both the renal cortex and the medulla renis prior to the
onset of recirculation. Since the tracer input is introduced into the renal artery,
it is distributed between compartments 1 and 2 with unknown fractional coefficients
a and 1- a. The major portion of the blood is passing to the cortex to be filtered;
hence we will suppose $1/2 < a < 1$. Moreover, it is then reasonable to assume that
the output from the cortex is larger than that from the medulla; thus we take
$a_{01} > a_{02} > 0$ in the model. Therefore the matrices of the system are given by

$$A = \begin{bmatrix} -a_{01} & 0 \\ 0 & -a_{02} \end{bmatrix} \qquad B = \begin{bmatrix} a \\ 1-a \end{bmatrix} \qquad C = [1 \quad 1] .$$

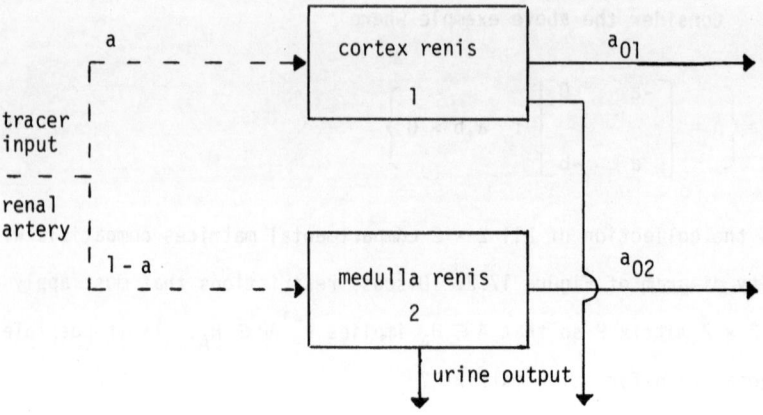

Figure 21.1

Two Compartment Model of Kidney Blood Flow

Using the theory developed in this section, the system is CC and CO since a_{01} is distinct from a_{02} and $a \neq 0, 1$. The following classes are assigned for the candidate matrices A, B, C from a priori knowledge regarding system structure:

$$H_A \equiv \left\{ \begin{bmatrix} -b & 0 \\ 0 & -c \end{bmatrix} : b > c > 0 \right\}$$

$$H_B \equiv \left\{ \begin{bmatrix} a \\ 1-a \end{bmatrix} : 1/2 < a < 1 \right\}$$

$$H_C \equiv \{[1 \quad 1]\} .$$

For $A \in H_A$, $B \in H_B$, and $C \in H_C$, we investigate the transformations $P^{-1}AP$, $P^{-1}B$, and CP to observe how $P \equiv [p_{ij}]$ is restricted with regard to the three classes. The condition $CP \in H_C$ demands that $CP = C$ which results in

(21.12) $P_{11} = 1 - P_{21}$, $P_{22} = 1 - P_{12}$.

The requirement $P^{-1}AP \in H_A$ implies that $P^{-1}AP = A'$ or $AP = PA'$, where $A' \in H_A$ has the form

$$\begin{bmatrix} -b' & 0 \\ 0 & -c' \end{bmatrix} \qquad b' > c' > 0.$$

Equating the four corresponding entries of AP and PA' yields the equations

$$b(1 - P_{21}) = b'(1 - P_{21})$$

(21.13)

$$b\, P_{12} = c'\, P_{12}$$

$$c\, P_{21} = b'\, P_{21}$$

$$c(1 - P_{12}) = c'(1 - P_{12}) .$$

By way of contradiction, let us assume P_{12} is nonzero in the second equation of (21.13). Then $b = c'$ and the fourth equation of (21.13) then says that $b = c$ or $P_{12} = 1$. The former cannot be the case since $b > c$ in this model. If $P_{12} = 1$, then $P_{22} = 0$ by (21.12). So now we are dealing with the matrix form

$$(21.14) \qquad P = \begin{bmatrix} 1 - P_{21} & 1 \\ P_{21} & 0 \end{bmatrix} .$$

The condition $P^{-1}B \in H_B$ or $B = PB'$ for B, $B' \in H_B$ and P given in (21.14) then leads to the restriction

$$(21.15) \qquad 1 - a = P_{21} a' .$$

Since $a \neq 1$ and $a' \neq 0$, condition (21.15) means that P_{21} cannot be zero. Here the third equation of (21.13) yields $c = b'$. Substituting c for b' in the first equation of (21.13) gives $(b - c)(1 - P_{21}) = 0$. Because $b \neq c$, then P_{21} must be unity. However, this cannot happen since by (21.15), there obtains $1 - a = a'$, and this relationship cannot hold due to the model restriction $1/2 < a, a' < 1$. Hence $P_{12} = 0$.

It now follows from (21.12) that $p_{22} = 1$. If $p_{21} \neq 0$, then the argument starting in the middle of the last paragraph leads to a contradiction. Hence $p_{21} = 0$ and then $p_{11} = 1$. This fully confirms that P is the identity matrix and thus the system (A, B, C) is structurally identifiable. This example demonstrates the following theorem of Travis et al. [228], [227].

THEOREM 21.5.   Assume the compartmental system is CC and CO with zero initial condition, and that H_A, H_B, and H_C are prescribed classes for the matrices A, B, and C, respectively. Then the compartmental system (A, B, C) is structurally identifiable if and only if $P^{-1}AP \in H_A$, $P^{-1}B \in H_B$, and $CP \in H_C$ together imply P = I.

Exercise 21.7.   Using Theorem 21.5, investigate the structural identification of (A, B, C), where H_A is given in Exercise 21.6, and H_B, H_C are specified as singletons containing only $B = \begin{bmatrix} 1 & 0 \end{bmatrix}^T$ and $C = \begin{bmatrix} 0 & 1 \end{bmatrix}$, respectively.

An interesting observation that comes out of Theorem 21.5, and is demonstrated in the last example, is that structural identification in the case of a CC and CO system is theoretically equivalent to solving for the entries of the matrix P, and from the conditions that are listed, this is seen to be a linear problem. However, the problem of arriving at the model parameter estimates given knowledge of the transfer function and its coefficients yet remains a nonlinear problem. For instance, in the above example, which we now know is structurally identifiable, the actual equations for estimation of the model parameters from the known transfer function

$$\hat{\phi}(s) = (s + \gamma_1)/(s^2 + s_1 s + s_2)$$

are

$$s_1 = a_{01} + a_{02}$$

(21.16) $$s_2 = a_{01} a_{02}$$

$$\gamma_1 = (1 - a)a_{01} + a \, a_{02} \; .$$

The first two equations are symmetric in a_{01} and a_{02} thereby yielding two solutions for a_{01} and a_{02}. However, only the single solution (a_{01}, a_{02}) with $a_{01} > a_{02} > 0$ falls within the feasible region dictated by the requirements of the model. From (21.16) it is noticed that the square root

$$(s_1^2 - 4s_2)^{\frac{1}{2}} = a_{01} - a_{02} > 0.$$

This relationship, combined with the first equation of (21.16), gives

$$a_{01} = [s_1 + (s_1^2 - 4s_2)^{\frac{1}{2}}]/2$$

$$a_{02} = s_1 - a_{01} .$$

With the excretions known, there follows

$$a = (\gamma_1 - a_{01})/(a_{02} - a_{01}),$$

the final parameter identification.

We close this section by mentioning one other result of Travis et al. [228], [227].

THEOREM 21.6.   Assume the compartmental system is either CC or CO and that the number of compartments is fixed. Then the system is structurally identifiable provided the conditions $P^{-1}AP \in H_A$, $P^{-1}B \in H_B$, and $CP \in H_C$ imply $P = I$.

Similarity transformations and sets of equivalent models have been more fully treated in [228], [227], [237], [231]. The methods were somewhat anticipated in the early work of Berman and Schoenfeld [31].

SECTION 22. MODEL IDENTIFICATION FROM THE TRANSFER FUNCTION EQUATIONS

22A. Form of the Nonlinear Equations

In Sections 16 through 21 we investigated the compartmental model identification problem from a number of different directions. Here we are concerned with
the particular system of nonlinear algebraic equations used for the identification
of the positive fractional transfer coefficients a_{ij} that arise from the transfer
function of the system (A, B, C). The general form that these equations take was
hinted at in Section 19, with concrete examples in (19.5), (19.8), and (19.15).
The derivation of that form is given presently. Once we have the form of these
equations well in mind, then simple necessary conditions for the existence of positive solutions to the equations can be presented.

In this section, let us assume that in the system (A, B, C), the matrices B
and C are known and are in fact (0, 1)-matrices (i.e., they only contain zeros and
ones). As we have previously discussed, the identification problem of finding the
unknown a_{ij}'s from experimental data becomes one of identifying A from knowledge of
the transfer function $\hat{\phi}(s) = C(sI - A)^{-1}B$. For a particular experiment (selection
of B and C), this relationship will provide the set (17.4) and (17.5) of nonlinear
equations in the unknown a_{ij}s (i ≠ j) in terms of the known transfer function coefficients r_k and s_m of (16.17). Our desire here is to describe the form that these
coefficients r_k and s_m must take in terms of the a_{ij} of the model. Example nonlinear systems (19.5), (19.8), and (19.15) suggest that the unknowns α_j (which
represents a particular a_{km} (k ≠ m)) come in a collection of equations, each
equation of which (a) has terms with coefficients of unity; (b) has terms that are
products of certain unknowns among α_1, α_2, ..., α_ν; (c) has the same number of
factors per term. Moreover, the number of factors in each term of matrix entry
$(r_m)_{k\ell}$, or the characteristic polynomial coefficient s_m, is m. We will find that
these statements remain valid for the general case of the coefficients in $\hat{\phi}$.

22B. Coefficients of the Transfer Function

We begin with the coefficients of the characteristic polynomial (16.17) (also
see [41]).

THEOREM 22.1.   Each s_m (m = 1, 2, ..., n) is the sum of products, each product of the type

$$a_{i_1 i_2} \, a_{i_3 i_4} \, \cdots \, a_{i_{2m-1} i_{2m}}$$

with m distinct factors, where the nonnegative integers satisfy $i_1 \neq i_2$, $i_3 \neq i_4$, ..., $i_{2m-1} \neq i_{2m}$.

Proof.   Let m be arbitrary but fixed in the set {1, 2, ..., n}.   As we know from Section 16, the scalar s_m is $(-1)^m$ times the sum of all m-square principle minors of A.   Consider any such principle minor:

$$A(i_1, i_2, \ldots, i_m) \equiv \begin{bmatrix} a_{i_1 i_1} & a_{i_1 i_2} & \cdots & a_{i_1 i_m} \\ a_{i_2 i_1} & a_{i_2 i_2} & \cdots & a_{i_2 i_m} \\ \cdots\cdots & \cdots\cdots & \cdots & \cdots\cdots \\ a_{i_m i_1} & a_{i_m i_2} & \cdots & a_{i_m i_m} \end{bmatrix}.$$

Define the m × m matrix $E \equiv [e_{k\ell}]$ by $E \equiv A(i_1, \ldots, i_m)$.   The evaluation of the minor is by

(22.1) $\det E = \sum\limits_{\rho} \varepsilon(j_1, j_2, \ldots, j_m) e_{1j_1} e_{2j_2} \cdots e_{mj_m}$

where the summation extends over all ρ = m! permutations $j_1, \ldots, j_m$ of the integers 1, ..., m.   The value of ε is 1 if its associated permutation is even; otherwise the value is -1.   In the matrix $A(i_1, \ldots, i_m)$, we can replace each diagonal element $a_{i_j i_j}$ by

(22.2) $-\sum\limits_{\substack{k=0 \\ k=i_j}}^{n} a_{k i_j}$ j = 1, 2, ..., m

since A is a compartmental matrix.   This means that in (22.1), we will be consider-ing e_{ij_j}s where $i = j_i$.   Now the permutation

$$\begin{pmatrix} 1 & 2 & \cdots & i-1 & i & i+1 & \cdots & m \\ j_1 & j_2 & \cdots & j_{i-1} & i & j_{i+1} & \cdots & j_m \end{pmatrix}$$

can be written as a cycle of $m - 1$ symbols (omitting $i = j_i$) and in turn such a cycle is the product of $m - 2$ transitions. In (22.1), let p_k be the number of elements e_{ij_i} in term $k \geq 2$ where $i = j_i$. Here $0 \leq p_k < m$. In addition, we specify that the first term in (22.1) is to be $\varepsilon(1, 2, \ldots, m)e_{11}e_{22} \cdots e_{mm}$, and so $p_1 \equiv m$. Now the kth term ($k \geq 2$) of (22.1) contains p_k factors e_{rr} with the same first and second subscript. Thus, as above, the associated permutation for that term can be expressed as a cycle of $m - p_k$ symbols (omitting as symbols the subscripts of these p_k factors), and this cycle is the product of $m - p_k - 1$ transitions. All of this tells us we can express $\varepsilon(1, 2, \ldots, m) = 1$ in the first term of $\det E$, and

$$\varepsilon(j_1, j_2, \ldots, j_m) = (-1)^{m-p_k-1}$$

for the kth term ($k \geq 2$) of $\det E$. Hence (22.1) can be written as

$$(22.3) \qquad e_{11}e_{22} \cdots e_{mm} + \sum_{k=2}^{m!} (-1)^{m-p_k-1} z_k$$

where z_k is a product of m e_{rs}s, as in (22.1). Now in z_k we have replaced each $e_{jj} \equiv a_{i_j i_j}$ by the sum in (22.2). There are p_k such replacements in z_k. Due to the negative sign in (22.2), we can factor a $(-1)^{p_k}$ out of z_k and rewrite $z_k \equiv (-1)^{p_k} w_k$. The expression w_k is a sum of terms, each term being the product of m distinct factors a_{rs} from $A(i_1, \ldots, i_m)$, with r and s different. Hence (22.3) becomes

$$(22.4) \qquad e_{11}e_{22} \cdots e_{mm} + (-1)^{m-1} \sum_{k=2}^{m!} w_k.$$

In the first term of (22.4), we have replaced each of the factors $e_{jj} = a_{i_j i_j}$ by its associated column sum in A as indicated in (22.2). This yields a first term in (22.4) that can be written as $(-1)^{p_1} w_1$, where w_1 is the sum of all possible combinations of products of off-diagonal elements a_{rs} with r selected from $0, 1, \ldots,$ $s-1, s+1, \ldots, n,$ and s among the integers $i_1, i_2, \ldots, i_m$. Here each product has exactly m distinct factors. Recalling that p_1 = m, then (22.4) is the same as

(22.5) $(- 1)^m w_1 + (- 1)^{m-1} \sum_{k=2}^{m!} w_k .$

Now s_m is $(- 1)^m$ times the sum, each term of which is of form (22.5). Thus,
by (22.5), each term in s_m will itself be of the form

(22.6) $(- 1)^{2m} \left[w_1 - \sum_{k=2}^{m!} w_k \right] .$

Now add together all of these terms (22.6) to get s_m. The constant $(- 1)^{2m}$ = 1 is
the coefficient of each term. The sum of all of the expressions w_1 will contain all
of the terms in the sum of the expressions

$- \sum_{k=2}^{m!} w_k ,$

but with some additional terms. The resulting cancellation of terms shows that s_m
is the sum of those additional terms, each of which has the form indicated in the
statement of the Theorem.

$\boxed{/\!/\!/}$

We now investigate the structure of the coefficients Γ_m in the numerator of $\hat{\phi}$.
By (17.4) we see that these matrix coefficients involve the s_m, and so the form of
s_m as indicated in Theorem 22.1 will aid in describing Γ_m. Since Γ_m is expressible
as (m = 0, 1, ..., n - 1)

(22.7) $C[A^m + s_1 A^{m-1} + ... + s_m I]B$

and because C and B are (0, 1)-matrices, we need only treat the matrices inside the
square brackets [·] of (22.7).

Let us start with the most elementary case m = 1; thus we consider the matrix
$A + s_1 I$. The general (i, j)-element of this matrix is a_{ij} if i $\neq$ j, and $a_{jj} + s_1$ if
i = j. In the second instance the entry is

$$a_{jj} + s_1 = a_{jj} - \sum_{k=1}^{n} a_{kk}$$

$$= - \sum_{\substack{k=1 \\ k \neq j}}^{n} a_{kk} = - \sum_{\substack{k=1 \\ k \neq j}}^{n} \left(- \sum_{\substack{m=0 \\ m \neq k}}^{n} a_{mk} \right).$$

Thus we see that each element of $A + s_1 I$ is a sum of singletons a_{rs} where $r \neq s$.

Due to the form derived for s_{m+1} in Theorem 22.1, we are lead to the general case via Mason's rule [41], [87, Chapter 8], [178] for signal flow graphs.

THEOREM 22.2.   In the transfer function (17.2), the general element of each numerator coefficient matrix Γ_m of (17.4) is of the form which is the sum of products

$$a_{i_1 i_2} \, a_{i_3 i_4} \, \cdots \, a_{i_{2m-1} i_{2m}}$$

with m distinct factors, and where $i_1 \neq i_2$, $i_3 \neq i_4$, $\ldots$, $i_{2m-1} \neq i_{2m}$.

22C. The Nonlinear Algebraic Equations for the Identification Problem

The purpose of the above material in this section is to demonstrate that the nonlinear equations arising in the identification of the system matrix A entries through the use of the transfer function for (A, B, C), when B and C are (0, 1)-matrices, are of the form $F(\alpha) = 0$ where $\alpha \equiv (\alpha_1, \alpha_2, \ldots, \alpha_\nu)^T$. Here α should be in the positive orthant of R^ν and the ith component function of F is

$$(22.8) \qquad F_i(\alpha) \equiv \sum_{j=1}^{p_i} \prod_{k=1}^{k_i} \alpha_{q(i,j,k)} - c_i$$

$i = 1, 2, \ldots, \nu$. Each $q(i, j, k)$ is a distinct integer selected from the set $\{1, 2, \ldots, \nu\}$. The integer p_i is positive, and the c_i are known real (in fact, positive) numbers (which are the s_m or the entries of Γ_m in $\hat{\phi}$). The integer k_i, $1 \leq k_i \leq \nu$, deserves special comment. For any given equation i of (22.8), k_i is of course specified in advance. However, each term of that equation has exactly k_i factors α_q and so k_i is independent of the term number j. This aspect of the form of equations (22.8) is a direct consequence of Theorems 22.1 and 22.2, and the fact that B and C of the system (A, B, C) are (0, 1)-matrices.

As an example, the equations (19.5) written in the form of (22.8) in order of increasing nonlinearity (increasing k_i) are

$$\alpha_1 + \alpha_2 + \alpha_3 + \alpha_4 + \alpha_5 - c_1 = 0 \qquad\qquad P_1 = 5, \ k_1 = 1$$

$$\alpha_3 + \alpha_4 + \alpha_5 - c_2 = 0 \qquad\qquad P_2 = 3, \ k_2 = 1$$

$$(22.9) \quad \alpha_2\alpha_3 + \alpha_1\alpha_4 + \alpha_2\alpha_4 + \alpha_1\alpha_5 + \alpha_2\alpha_5 + \alpha_3\alpha_5 - c_3 = 0 \qquad\qquad P_3 = 6, \ k_3 = 2$$

$$\alpha_3\alpha_5 - c_4 = 0 \qquad\qquad P_4 = 1, \ k_4 = 2$$

$$\alpha_2\alpha_3\alpha_5 - c_5 = 0 \qquad\qquad P_5 = 1, \ k_5 = 3$$

(Here, α_3 and α_4 have been reversed from that in 19.5.)

It is important to note that since

$$s_1 = (-1) \sum_{m=1}^{n} a_{mm} = (-1) \sum_{m=1}^{n} \left[- \sum_{\substack{p=0 \\ p \neq m}}^{n} a_{pm} \right] ,$$

which is the sum of all the unknowns a_{ij} ($i \neq j$) in the compartmental matrix A, we will anticipate in (22.8) the linear form

$$(22.10) \qquad F_1(\alpha) \equiv \alpha_1 + \alpha_2 + \ldots + \alpha_\nu - c_1 = 0$$

as the leading equation of the system $F(\alpha) = 0$, and the only linear equation involving every unknown. This fact will be critical in establishing the necessary conditions on (22.8) yet to come.

22D. Necessary Conditions for Positive Solutions of the Nonlinear System

Now that we have placed the compartmental identification equations under the general classification as governed by (22.8), we can address the question of how to find necessary conditions for the existence of positive solutions of these equations.

The positivity and product nature of the terms in each F_i suggests the possibility of utilizing the arithmetic mean-geometric mean inequality

$$(\Delta_1 \Delta_2 \cdots \Delta_k)^{1/k} \leq (\Delta_1 + \Delta_2 + \dots + \Delta_k)/k$$

which holds for $\Delta_i \geq 0$ and k a positive integer.

THEOREM 22.3.   Suppose the first equation of system $F(\alpha) = 0$ is (22.10). If there exists a positive solution $\alpha \equiv (\alpha_1, \alpha_2, \dots, \alpha_\nu)^T$ of $F(\alpha) = 0$, then the constants c_i must conform to the conditions

$$(22.11) \qquad 0 < c_i < p_i(c_1/k_i)^{k_i} \qquad\qquad i = 2, 3, \dots, \nu.$$

Proof.   Because of the existence of a positive solution α of $F(\alpha) = 0$,

$$(22.12) \qquad 0 < c_i = \sum_{j=1}^{p_i} \alpha_{q(i,j,1)} \cdots \alpha_{q(i,j,k_i)}$$

holds for each $i = 2, 3, \dots, \nu$. Since all components $\alpha_1, \dots, \alpha_\nu$ of α are positive, the arithmetic mean-geometric mean inequality applies, and we get

$$\left[\alpha_{q(i,j,1)} \cdots \alpha_{q(i,j,k_i)}\right]^{1/k_i} \leq \frac{1}{k_i} \sum_{k=1}^{k_i} \alpha_{q(i,j,k)} \cdot$$

Hence the sum in (22.12) is equal to

$$\sum_{j=1}^{p_i} \left(\left[\alpha_{q(i,j,1)} \cdots \alpha_{q(i,j,k_i)}\right]^{1/k_i}\right)^{k_i}$$

$$(22.13) \qquad\qquad \leq \sum_{j=1}^{p_i} \left(\frac{1}{k_i} \sum_{k=1}^{k_i} \alpha_{q(i,j,k)}\right)^{k_i} \cdot$$

Now the inner sum of (22.13) is a partial sum of all the variables α_m, and since these variables are nonnegative that partial sum is less than or equal to the sum $\alpha_1 + \alpha_2 + \dots + \alpha_\nu$ of all the variables. Moreover, according to hypothesis condition (22.10), that sum is c_1. Thus (22.13) is less than

$$(22.14) \qquad \sum_{j=1}^{p_i} (c_1/k_i)^{k_i} = p_i (c_1/k_i)^{k_i},$$

the last equality due to the fact that k_i is the same for all j. Combining (22.12)

through (22.14), we arrive at the stated inequality (22.11).

⁄⁄⁄

These necessary conditions are very simple to apply, and we note that they are

really conditions on the entries of Γ_m and the scalars s_m of the transfer function,

in terms of s_1. As an example, the $\nu = 5$ equations (22.9) for the catenary model

with single exit from the first compartment yield the following necessary conditions

on the c_i:

$$0 < c_2 < 3c_1 \qquad\qquad 0 < c_4 < (c_1/2)^2$$

$$0 < c_3 < 6(c_1/2)^2 \qquad\qquad 0 < c_5 < (c_1/3)^3 .$$

The reason why inequalities such as these are of interest is because they provide

useful eliminators of otherwise plausible nonlinear equations. This is illustrated

by another example. Consider the general equations of (19.15),

$$\alpha_1 + \alpha_2 + \alpha_3 - c_1 = 0$$

$$(22.15) \qquad \alpha_1\alpha_2 + \alpha_2\alpha_3 - c_2 = 0$$

$$\alpha_1\alpha_2\alpha_3 - c_3 = 0.$$

Suppose $c_1 = 1.40$, $c_2 = 0.713$, and $c_3 = 0.121$ in (22.15). The result of Theorem

22.3 is that if there is a solution α_1, α_2, $\alpha_3 > 0$ to these nonlinear equations,

then we must have

$$(22.16) \qquad 0 < c_2 < 2(c_1/2)^2, \qquad\qquad 0 < c_3 < (c_1/3)^3.$$

The former inequality is satisfied, but the latter is not. Hence we must conclude

that there does not exist a positive (in fact, nonnegative) solution of this system.

Equations (22.15) were further analyzed using subroutine ZSYSTM in the IMSL package

[131]. This nonlinear algebraic equation solver was designed by Brown [46] and is

especially applicable to equations (22.8) if they are arranged in order of increasing nonlinearity. This is due to the design (involving linearization by steps through the set of equations) of the algorithm. In the case of (22.15) with $c_1 = 1.40$, $c_2 = 0.713$, and $c_3 = 0.121$, that no positive solution existed was confirmed numerically using ZSYSTM. Moreover, numerical experiments such as increasing c_1 to the value 1.5 and then 1.6 (so that both inequalities of (22.16) are satisfied) still presented a system with no apparent positive solution. However, moving c_1 up to the value 1.69 as in (19.15) gave the locally unique positive solution $(0.64697, 0.81298, 0.23005)^T$.

Exercise 22.1.   Consider system (22.15) with $c_1 = 6$, $c_2 = 8$, and $c_3 = 6$. Do conditions (22.11) apply? How many positive solutions are there?

Exercise 22.2.   If $c_1 = 17$, $c_2 = 13$, $c_3 = 53$, $c_4 = 2$, and $c_5 = 6$ in (22.9), what do the inequalities (22.11) tell us? Find a positive solution of the nonlinear system. Is it unique?

22E.   Refined Necessary Conditions

If some equation $i \geq 2$ of system (22.8) is <u>linear</u> in all its variables (e.g., the second equation of (22.9)), then the corresponding inequality of (22.11) can be easily tightened. This is because if a positive solution $\alpha = (\alpha_1, \ldots, \alpha_\nu)^T$ of $F(\alpha) = 0$ exists, then under hypothesis condition (22.10) of Theorem 22.3, and since $k_i = 1$,

$$c_i = \sum_{j=1}^{p_i} \alpha_{q(i,j,1)} < \alpha_1 + \alpha_2 + \ldots + \alpha_\nu = c_1 .$$

Moreover, it is clear that other necessary conditions can be derived if some $k_i = 1$ for $i \geq 2$.

We can also improve conditions (22.11) in another way. For a given i, inequality (22.11) gets weaker as p_i becomes larger, that is, when the number of terms in that equation increase. We look for an inequality that strengthens (22.11) in such cases. Consider, for $i = 2, 3, \ldots, \nu$,

$$c_i^{1/k_i} = \left[\sum_{j=1}^{p_i} \alpha_{q(i,j,1)} \cdots \alpha_{q(i,j,k_i)} \right]^{1/k_i}$$

$$< \sum_{j=1}^{p_i} \prod_{k=1}^{k_i} \alpha_{q(i,j,k)}^{1/k_i} \leq \prod_{k=1}^{k_i} \left[\sum_{j=1}^{p_i} \alpha_{q(i,j,k)} \right]^{1/k_i} ,$$

where the last inequality is valid by problem H, p. 193 of [189] and because k_i is independent of j. Under hypothesis (22.10) the last product above is less than

$$\prod_{k=1}^{k_i} [\alpha_1 + \alpha_2 + \dots + \alpha_\nu]^{1/k_i} = \prod_{k=1}^{k_i} [c_1]^{1/k_i} = c_1 .$$

When the number of terms p_i in equation $i \geq 2$ satisfies the additional condition $p_i > k_i^{k_i}$, then inequality

(22.17) $c_i < c_1^{k_i}$

implies that

$$c_i < c_1^{k_i} = k_i^{k_i} (c_1/k_i)^{k_i} < p_i (c_1/k_i)^{k_i}$$

so that (22.17) is stricter than (22.11). Thus in summary we have the next statement which supersedes Theorem 22.3.

THEOREM 22.4.   Let the first equation of system (22.8) be the full linear equation (22.10), and suppose there exists a positive solution of the system. If equation $i \geq 2$ of (22.8) is

(a)   linear,   then $0 < c_i < c_1$;

(b)   such that $p_i \leq k_i^{k_i}$,   then $0 < c_i < p_i (c_1/k_i)^{k_i}$;

(c)   such that $p_i > k_i^{k_i}$,   then $0 < c_i < c_1^{k_i}$.

Exercise 22.3. Apply the conditions of Theorem 22.4 to system (22.9). State explicitly any improvement over inequalities (22.11). Compare with the results of Exercise 22.2.

Exercise 22.4. The identification of model parameters that arise in a moving average process or in ARMA models requires the solution of system (22.8) when $p_i = \nu - i + 1$, $k_i = 2$ for all i (thus all equations are "quadratic"), $q(i,j,1) = j$, and $q(i,j,2) = i + j - 1$, so that the system (22.8) becomes

$$(22.18) \qquad F_i(\alpha) \equiv \sum_{j=1}^{\nu-i+1} \alpha_j \alpha_{i+j-1} - c_i = 0$$

$i = i, 2, \ldots, \nu$ [10], [11], [242]. By using similar tricks involving the arithmetic mean-geometric mean inequality as were employed above, find necessary conditions on the c_is of (22.18) for real solutions of the system $F(\alpha) = 0$ to exist. Conditions on c_i can also be found using the Rayleigh Principle. Also see [11], [13], [73].

22F. Additional Properties of the Nonlinear Algebraic System

Along with the above necessary conditions, we can derive a uniqueness result for the system $F(\alpha) = 0$ in the positive orthant. Let us suppose, in $F(\alpha) = 0$, that the equations are such that equation i (for each i = 1, 2, ..., ν) necessarily contains variable α_i, and otherwise only variables α_j with $j \geq i$ (this may require some renumbering of variables and/or reordering of equations). Then the system can be solved for α by back-substitution starting with the last variable α_ν. Thus the following result obtains.

THEOREM 22.5. If the ith equation (i = 1, 2, ..., ν) of system $F(\alpha) = 0$ as given in (22.8) contains variable α_i and otherwise only involves variables α_j with $j \geq i$, and if there exists a positive solution of $F(\alpha) = 0$, then that solution is unique.

An example of this is the set of equations (16.11) arising from the two compartment system in Figure 16.1 when there is no excretion to the outside environment from the second compartment (i.e., $a_{02} = 0$). Another illustration is the system

$$\alpha_1 + \alpha_2 + \alpha_3 + \alpha_4 - c_1 = 0$$

$$\alpha_2\alpha_3 + \alpha_2\alpha_4 - c_2 = 0$$

$$\alpha_3\alpha_4 - c_3 = 0$$

$$\alpha_4 - c_4 = 0 .$$

The homogeneity within each equation in (22.8) of nonlinear system $F(\alpha) = 0$ leads to the fact that the Jacobian $F'(\alpha)$ must be a polynomial homogeneous in the variables $\alpha_1, \alpha_2, \ldots, \alpha_\nu$. Thus for all scalars $\lambda > 0$, and for all $\alpha \equiv (\alpha_1, \alpha_2, \ldots, \alpha_\nu)^T > 0$, there exists some positive integer μ such that

(22.19) $\det F'(\lambda\alpha) = \lambda^\mu \cdot \det F'(\alpha).$

Discovering a formula for μ merits further work. Condition (22.19) on the αs says that if $\det F'(\alpha)$ is nonzero at some point $\alpha = \alpha^0$, then there is a "cone" projecting out from the origin that contains α^0 as an interior point such that the Jacobian $F'(\alpha)$ is a nonsingular matrix for all α in this cone. This nice situation suggests that we need only look at those positive α-vectors in R^ν that have norm $\| \alpha \| = \gamma$ for some choice $\| \cdot \|$ of norm and for some $\gamma > 0$. However, this may not tell us much more than we already know, for if we choose the 1-norm and $\gamma \equiv c_1 > 0$, the "sphere" $\| \alpha \|_1 = c_1$ is just the first equation (22.10) of $F(\alpha) = 0$:

$$\sum_{i=1}^{\nu} \alpha_i = \sum_{i=1}^{\nu} |\alpha_i| \equiv \| \alpha \|_1 = c_1 \equiv \gamma ,$$

restricting ourselves to the positive orthant of R^ν.

Exercise 22.5.   Compute the determinant of the Jacobian $F'(\alpha)$ of system (22.15) and verify that it is homogeneous in the variables α_i. Find μ of (22.19) and also find regions in the positive orthant of R^3 on which $F'(\alpha)$ is singular. Determine at least two solutions of $F(\alpha) = 0$ in this orthant.

22G. An Iterative Scheme for Solving F(α) = 0

Iterative methods for solving $F(\alpha) = 0$ were discussed in Section 19 (e.g., (19.2), (19.7)). Now that we have determined an explicit form for these equations, namely (22.8), we can present one other iterative scheme with which we have had some success. Consider, as a concrete illustration, equations (22.15) again. With $\alpha > 0$, let α_1 represent the smallest of the three variables α_1, α_2, α_3. Divide both sides of equation i by α_1^{ki}, $i = 1, 2, 3$ to get the system

$$1 + (\alpha_2/\alpha_1) + (\alpha_3/\alpha_1) = c_1/\alpha_1$$

$$(\alpha_2/\alpha_1) + (\alpha_2/\alpha_1)(\alpha_3/\alpha_1) = c_2/\alpha_1^2$$

$$(\alpha_2/\alpha_1)(\alpha_3/\alpha_1) = c_3/\alpha_1^3 .$$

Rewrite this set of equations as

$$\alpha_1 = c_1/[1 + (\alpha_2/\alpha_1) + (\alpha_3/\alpha_1)]$$

$$(\alpha_2/\alpha_1) = (c_2/\alpha_1^2) - (\alpha_2/\alpha_1)(\alpha_3/\alpha_1)$$

$$(\alpha_3/\alpha_1) = (c_3/\alpha_1^3)/(\alpha_2/\alpha_1).$$

Now define $t_1 \equiv \alpha_2/\alpha_1$ and $t_2 \equiv \alpha_3/\alpha_1$. Then we use the algorithm

$$\alpha_1 = c_1/[1 + t_1^{(m)} + t_2^{(m)}]$$

$$(22.20) \qquad t_1^{(m+1)} = (c_2/\alpha_1^2) - t_1^{(m)} t_2^{(m)}$$

$$t_2^{(m+1)} = (c_3/\alpha_1^3)/t_1^{(m+1)}$$

with <u>updating</u> of t_1 in the third equation. Algorithms constructed in an analogous manner have been successful in dealing with a number of different equations of type (22.8).

<u>Exercise 22.6.</u> With $c_1 = 6$, $c_2 = 8$, and $c_3 = 6$, and using initial estimates of $t_1^0 = 1.8$, $t_2^0 = 3.1$, show that algorithm (22.20) yields a sequence which converges

to $(t_1^*, t_2^*) = (2, 3)$, from which we can determine the solution $\alpha^* = (1, 2, 3)$. Try to set up a proof of convergence of (22.20) through the use of the Contractive Mapping Theorem [40]. Discuss existence and uniqueness of solution.

22H. Triangularization of $F(\alpha) = 0$

We can outline steps of a general procedure that are useful in treating equations (22.8). The basic purpose here is to take the n-variable system $F(\alpha) = 0$ and transform it into a <u>triangular system</u> $T(\phi) = 0$ of form

$$T_1(\phi_1) = 0$$

$$T_2(\phi_1, \phi_2) = 0$$

(22.21)

$$\vdots$$

$$T_n(\phi_1, \phi_2, \ldots, \phi_n) = 0$$

where $\phi_1, \ldots, \phi_n$ is just a renumbering of the variables $\alpha_1, \ldots, \alpha_n$. Then (22.21) is solved by forward substitution to get a solution of $F(\alpha) = 0$. We attempt to accomplish this triangularization through the following kinds of steps:

Step 1. Make any obvious replacements of variable sums or products by the appropriate constants c_i, but if possible avoid writing a division by a variable (so as to not introduce extraneous roots, and to keep expressions in polynomial form).

Let us illustrate with system (22.9). After replacement, the first, third, and fifth equations of this system become

$$\alpha_1 + \alpha_2 + c_2 = c_1$$

(22.22a)

$$\alpha_2\alpha_3 + \alpha_1\alpha_4 + \alpha_2\alpha_4 + \alpha_1\alpha_5 + \alpha_2\alpha_5 + c_4 = c_3$$

$$\alpha_2 c_4 = c_5$$

and we still have the other equations

$$\alpha_3 + \alpha_4 + \alpha_5 = c_2$$

(22.22b)

$$\alpha_3\alpha_5 = c_4.$$

This process can have the effect of reducing the degree of the equations. For instance, in (22.9) the largest k_i is three, however in (22.22) the highest k_i is two.

Step 2.   Starting with the most nonlinear equations (highest k_i), replace variables by the same variables from some <u>linear</u> equation ($k_i = 1$).

For our example equations, there obtains

$$\alpha_2\alpha_3 + \alpha_1\alpha_4 + \alpha_2\alpha_4 + (\alpha_1 + \alpha_2)(c_2 - \alpha_3 - \alpha_4) + c_4 = c_3$$

(22.23)

$$\alpha_3(c_2 - \alpha_3 - \alpha_4) = c_4$$

with a substitution for α_5 from the linear equation of (22.22b). This substitution will generally allow some terms to cancel in the equations so treated.

Step 3.   Continue this process of reduction and elimination of variables to place the system in triangular form $T(\phi) = 0$.

In our illustration, equations (22.23) and the remaining equations of (22.22) can be written as

$$\alpha_2 c_4 = c_5$$

$$\alpha_1 + \alpha_2 + c_2 = c_1$$

(22.24)

$$c_2(\alpha_1 + \alpha_2) - \alpha_1\alpha_3 + c_4 = c_3$$

$$c_2\alpha_3 - \alpha_3^2 - \alpha_3\alpha_4 = c_4$$

$$\alpha_3 + \alpha_4 + \alpha_5 = c_2$$

System (22.24) is in the format of (22.21) with $\phi_1 \equiv \alpha_2$, $\phi_2 \equiv \alpha_1$ and $\phi_i \equiv \alpha_i$, $i = 3, 4, 5$. Each $T_i(\phi_1, \ldots, \phi_i)$ is a polynomial in the variables $\phi_1, \ldots, \phi_i$.

Thus by the Fundamental Theorem of Algebra, we know the actual number of roots of $T_1(\phi_1) = 0$, and then of $T_2(\phi_1, \phi_2) = 0$, etc., until finally all roots of $T(\phi) = 0$ (i.e., $F(\alpha) = 0$) can be listed. With our present example, we observe that the equations of (22.24) are successively <u>linear</u> in the unknowns in the order ϕ_1, ϕ_2, ..., ϕ_5, and thus we immediately see that there is a <u>unique</u> solution of system (22.24) and thus of system (22.9).

<u>Exercise 22.7.</u> Show that system (22.15) can be reduced to the triangular system

$$c_1\alpha_2 - \alpha_2^2 = c_2$$

$$c_1\alpha_1\alpha_2 - \alpha_1^2\alpha_2 - \alpha_1\alpha_2^2 = c_3$$

$$\alpha_1 + \alpha_2 + \alpha_3 = c_1 \;.$$

With $c_1 = 6$, $c_2 = 8$, and $c_3 = 6$, discuss all possible positive roots of the original system (22.15).

<u>Exercise 22.8.</u> For

$$\alpha_1 + \alpha_2 + \alpha_3 + \alpha_4 = c_1$$

$$\alpha_1 + \alpha_3 + \alpha_4 = c_2$$

$$\alpha_1\alpha_2 + \alpha_1\alpha_3 + \alpha_2\alpha_4 = c_3$$

$$\alpha_1\alpha_2\alpha_4 = c_4$$

verify that one can reduce this system to the triangular form

$$\alpha_2 + c_2 = c_1$$

$$c_3\alpha_1\alpha_2 - \alpha_2 c_4 = c_1\alpha_1^2\alpha_2 - \alpha_1^3\alpha_2 - c_4\alpha_1$$

$$\alpha_1\alpha_2\alpha_4 = c_4$$

$$\alpha_1 + \alpha_2 + \alpha_3 + \alpha_4 = c_1$$

Discuss the uniqueness of positive solutions of this system if c_1 = 10, c_2 = 8, c_3 = 13, and c_4 = 8.

There are presently efforts being made by certain investigators to study the handling of compartmental systems by symbolic manipulation using general algebraic computations software such as LISP and REDUCE 2. It will be interesting to see how symbolic manipulation might be effectively used for such things as deriving the nonlinear system $F(\alpha)$ = 0 from the compartmental model transfer function [41] and then reducing it to triangular form $T(\phi)$ = 0.

22I. Uniqueness of Solution of $F(\alpha)$ = 0

We close this section with some thoughts regarding the uniqueness of solution of $F(\alpha)$ = 0. These concepts need to be investigated further. Let

$$\Omega \equiv \{\alpha \equiv (\alpha_1, \ldots, \alpha_\nu) \in R^\nu : \text{ each } \alpha_i > 0\}$$

be the positive orthant in R^ν. Let M be a $\nu \times \nu$ real matrix such that $\alpha \in \Omega$ implies $M\alpha \in \Omega$. For the nonlinear function $F(\alpha)$ of (22.8), we look into the possibility that there exists an M such that $F(M\alpha)$ = $F(\alpha)$ for all α in Ω.

Exercise 22.9. Suppose, as usual, that the first function of (22.8) is $F_1(\alpha) \equiv \alpha_1 + \alpha_2 + \ldots + \alpha_\nu - c_1$ (k_1 = 1, p_1 = ν). Show that if each column sum of M is unity, then the first function of $F(M\alpha)$ has exactly the same structure. Are there any other conditions we can find on M so that $F(M\alpha)$ = $F(\alpha)$ will hold for all $\alpha \in \Omega$?

Now assume there exists a solution $\alpha^* \in \Omega$ of $F(\alpha)$ = 0. If in fact there is a matrix $M \neq I$ as discussed above, then $M\alpha^*$ is another solution distinct from α^* in Ω, and so the nonlinear system $F(\alpha)$ = 0 cannot have a unique positive solution. In particular, this may happen when M is a permutation matrix in which case we call the function F symmetric on Ω. An example of this is found in system (22.15). Here we discover that

$$M = \begin{bmatrix} 0 & 0 & 1 \\ 0 & 1 & 0 \\ 1 & 0 & 0 \end{bmatrix}$$

Then for all $\alpha \in \Omega$, $F(M\alpha) = F(\alpha)$, so that F is symmetric. Hence if α^* is a solution of $F(\alpha) = 0$, then also is $M\alpha^*$.

Another example to consider is the 3-variable system $F(\alpha) = 0$, with components

$$G_1(\alpha) \equiv \alpha_1 + \alpha_2 + \alpha_3 = c_1$$

(22.25) $\qquad G_2(\alpha) \equiv \alpha_1 \alpha_2 \qquad\qquad = c_2$

$$G_3(\alpha) \equiv \alpha_1 \alpha_3 \qquad\qquad = c_3 \; .$$

Suppose, when $c_1 = 6$, $c_2 = 2$, and $c_3 = 3$, we have computed by some numerical method the solution

(22.26) $\qquad \alpha^* = (1, 2, 3)^T$

in Ω. To investigate the possible uniqueness of this positive solution, we set up the transformation ($m_1 + m_2 + m_3 = 1$)

(22.27) $\qquad M\alpha = \begin{bmatrix} 1 & 0 & m_1 \\ 0 & 1 & m_2 \\ 0 & 0 & m_3 \end{bmatrix} \begin{bmatrix} \alpha_1 \\ \alpha_2 \\ \alpha_3 \end{bmatrix} = \begin{bmatrix} \alpha_1 + m_1\alpha_3 \\ \alpha_2 + m_2\alpha_3 \\ m_3\alpha_3 \end{bmatrix}$

in which use has been made of the fact that each column of M should sum to unity (see Exercise 22.9). We select $\nu = 3$ unknowns and restrict them to some column of M (say, the last column) since there are $\nu = 3$ equations in (22.27) to determine information. Now, using the vector $m \equiv (m_1, m_2, m_3)^T$, we compute $F(M\alpha^*)$. Its first component will always be zero since $G_1(m) = 1$. If $M\alpha^*$ is to be a solution of $F = 0$, then each of the remaining components of $F(M\alpha^*)$ must be zero. These two conditions determine a quadratic equation in the single unknown m_1, with solutions

$m_1 = 0$, 4/3. The value $m_1 = 0$ forces $m_2 = 0$ and $m_3 = 1$, so that $M = I$ and thus we are at the original computed solution (22.26). The value $m_1 = 4/3$ yields $m_3 = 1/5$, and $m_2 = -8/15$. Hence a second solution of $F(\alpha) = 0$ in Ω is

$$\alpha^{**} = M\alpha^* = \begin{bmatrix} 1 & 0 & 4/3 \\ 0 & 1 & -8/15 \\ 0 & 0 & 1/5 \end{bmatrix} \begin{bmatrix} 1 \\ 2 \\ 3 \end{bmatrix} \begin{bmatrix} 5 \\ 2/5 \\ 3/5 \end{bmatrix}.$$

Another observation (based on numerous examples) can be made about the uniqueness of solutions of $F(\alpha) = 0$. Suppose there exists a solution α^* in the positive orthant Ω for $F(\alpha) = 0$ (wherein all $c_i > 0$). It then appears that the following is true:

(22.28) α^* is the <u>only</u> solution of $F(\alpha) = 0$ if and only if either $\det F'(\alpha) > 0$ or $\det F'(\alpha) < 0$ for <u>all</u> $\alpha \in \Omega$.

Among the many examples studied, the 3-compartment model of Bossi et al. [41] gives rise to seven equations in five unknowns; of these, the system $F(\alpha) = 0$ consists of the independent equations

$$\alpha_1 + \alpha_2 + \alpha_3 + \alpha_4 + \alpha_5 - c_1 = 0$$

$$\alpha_1 + \alpha_4 + \alpha_5 - c_2 = 0$$

(22.29) $\alpha_3 - c_3 = 0$

$$\alpha_1\alpha_3 - c_4 = 0$$

$$\alpha_1\alpha_3\alpha_5 - c_5 = 0.$$

Here, $\det F'(\alpha) = \alpha_1\alpha_3^2$, which is positive over all of Ω. Moreover, the solution of (22.29) is unique. Another system $F(\alpha) = 0$ with unique solution is

$$\alpha_1 + \alpha_2 + \alpha_3 + \alpha_4 - c_1 = 0$$

$$\alpha_1 + \alpha_3 - c_2 = 0$$

(22.30)

$$\alpha_1\alpha_2 - c_3 = 0$$

$$\alpha_1\alpha_2\alpha_3 - c_4 = 0.$$

We find that $\det F'(\alpha) = -\alpha_1^2\alpha_2$, which is negative on the whole of Ω. On the other hand, the system of Exercise 22.8, with $c_1 = 10$, $c_2 = 8$, $c_3 = 13$, and $c_4 = 8$, possesses only two solutions,

$$\alpha^* = (1, 2, 3, 4,) \in \Omega$$

$$\alpha^{**} = (8, 2, -0.5, 0.5) \notin \Omega.$$

Computation of $\det F'(\alpha)$ yields

$$\alpha_1\alpha_2^2 + \alpha_1\alpha_2\alpha_3 + \alpha_1\alpha_2\alpha_4 - \alpha_2^2\alpha_4 - \alpha_1^2\alpha_2 ,$$

which can be zero in Ω.

SECTION 23. THE PARAMETER ESTIMATION PROBLEM

23A. The Basic Estimation Problem

We now turn to an important problem for the biomedical scientist of fitting an appropriate compartmental model to data that has been gathered from an experiment. At this point it is assumed that a linear compartmental model, based on known physiologic and anatomic considerations, has been constructed for the system under investigation. Moreover, we suppose that the design of the tracer experiment and the choice of data sampling locations has been determined, and the compartmental model has been shown to be (uniquely) structurally identifiable. Thus the problem now comes down to one of estimating parameters of the model, that is, of deriving all of the parameter values of the model from the available data.

The basic parameter estimation problem can be stated as follows. We have decided on an appropriate compartmental model (16.5) with components which are functions of parameter vector $\alpha \equiv (\alpha_1, \alpha_2, \ldots, \alpha_\nu)^T$, so that we rewrite (16.5) as

$$\dot{x}(t,\alpha) = A(\alpha)x(t,\alpha) + B(\alpha)u(t), \qquad t \geq 0$$

$$(23.1) \qquad x(0,\alpha) = 0$$

$$y(t,\alpha) = C(\alpha)x(t,\alpha)$$

for a particular tracer experiment which has been selected so as to ensure structural identification of the model. We wish to numerically determine by some criterion the unique model parameter vector α which is associated with the observed raw data that makes up $y(t,\alpha)$.

As an example, the renal cortex-medulla model considered at the close of Section 21 has matrices A and B of the form ($\alpha \equiv (\alpha_1, \alpha_2, \alpha_3)^T$)

$$(23.2) \qquad A(\alpha) \equiv \begin{bmatrix} -\alpha_1 & 0 \\ 0 & -\alpha_2 \end{bmatrix} \qquad B(\alpha) \equiv \begin{bmatrix} \alpha_3 \\ 1 - \alpha_3 \end{bmatrix}$$

and C does not depend on the parameters α_i since it is completely specified as [1 1].

The input $u(t)$ and output $y(t,\alpha)$ are related by

(23.3) $y(t,\alpha) = \phi(t,\alpha) * u(t)$

as we have discussed in Section 16. Thus, provided with the measured output $y(t,\alpha)$ and known input $u(t)$, we will assume that the impulse response function

(23.4) $\phi(t,\alpha) \equiv C(\alpha) \exp (tA(\alpha))B(\alpha)$

or equivalently, the transfer function

(23.5) $\hat{\phi}(s,\alpha) = C(\alpha)[sI - A(\alpha)]^{-1} B(\alpha)$

is known. Now the known coefficients of the transfer function are nonlinear functions of the unknown α: $c_i \equiv c_i(\alpha)$. These are the nonlinear equations of type (22.8) written as $F(\alpha) = 0$ that were investigated in Section 22. We wish to "invert" this system of equations and thus solve for α. Since structural identifiability of the compartmental system is assumed, that inversion is now possible, yielding a unique solution vector $\alpha^* > 0$. The discussion in this section will concentrate on that aspect of the problem which is determining y (and thus ϕ) from discrete experimental data by curve-fitting.

In standard analysis of most experiments utilizing tracer-labeled substances, the basic task in the parameter estimation problem is to convert the raw data, normally collected at discrete times (since $y(t)$ is usually "incomplete" due to rather severe biological restrictions), into the continuous representation $y(t)$. Due to the linear differential equation model (23.1), this representation consists of a "disappearance curve" $y(t)$ as a sum of exponentials

(23.6) $y(t) = A_1 \exp (\lambda_1 t) + A_2 \exp (\lambda_2 t) + \ldots + A_n \exp (\lambda_n t).$

Here we assume distinct eigenvalues λ_i of the $n \times n$ compartmental matrix A, and since we are fitting (23.6) to real data, that the λ_i are real and nonpositive:

$\lambda_1 < \lambda_2 < \ldots < \lambda_n \leq 0$. The output function parameters $\{A_i, \lambda_i\}$ all together constitute a real 2n-dimensional vector which will yield the vector c of transfer function coefficients of (23.5).

23B. A Lipoprotein Metabolism Model

To illustrate the above outline of events, let us consider the modeling of lipoprotein metabolism [37]. The single exit ($a_{02} \equiv 0$) system of Figure 16.1 is used where compartment 1 is the blood plasma, and compartment 2 is the extravascular space. The system matrix is

$$(23.7) \qquad A = \begin{bmatrix} a_{11} & a_{12} \\ a_{21} & a_{22} \end{bmatrix} \qquad \begin{array}{l} a_{11} = -a_{21} - a_{01} \\[2ex] a_{22} = -a_{12} \end{array}$$

and the experiment consists of a bolus injection of a unit amount of tracer into compartment 1 at time t = 0, and then the subsequent sampling of compartment 1 at times $t_1, t_2, \ldots, t_m$. Hence we take

$$u(t) = 1 \cdot \delta(t), \qquad B^T = [1 \quad 0] = C.$$

Thus all model parameters α_i fall within A and so we identify $\alpha_1 \equiv a_{21}$, $\alpha_2 \equiv a_{01}$, and $\alpha_3 \equiv a_{12}$ (also see (19.8)). In Exercise 16.2 we found that this model is structurally identifiable (this also follows from equations (19.8) and Theorem 22.5). Therefore there exists a unique vector $\alpha^* \equiv (\alpha_1^*, \alpha_2^*, \alpha_3^*)^T$ in the positive orthant

$$\Omega \equiv \{(\alpha_1, \alpha_2, \alpha_3) \in R^3 : \alpha_i > 0\}$$

that will give the solution to the problem.

We go about finding this α^* as follows. So that we have a concrete example in mind, consider the data set given in Table 23.1, which is based on baboon low-density-lipoprotein (LDL) data taken over a 10-day period in laboratory animal studies relating to the material in [37]. Let us assume for the moment that the least-squares fitting of this discrete data by the continuous curve

$$(23.8) \qquad y(t) = Cx(t) = x_1(t) \equiv A_1 e^{\lambda_1 t} + A_2 e^{\lambda_2 t}$$

has been accomplished, with the unique minimizing parameter values

(23.9)

$$A_1^* = 0.7539 \qquad A_2^* = 0.2454$$

$$\lambda_1^* = -2.1080 \qquad \lambda_2^* = -0.3989.$$

TABLE 23.1.

Baboon 170 LDL Data

Time t in days	Amount y(t) of tracer in compartment 1
0.00	1.0000
0.5	0.4610
1.0	0.2590
1.5	0.1700
2.0	0.1210
3.0	0.0722
4.0	0.0451
5.0	0.0319
6.0	0.0240
7.0	0.0182
8.0	0.0141
9.0	0.0100
10.0	0.0094

(This is the primary topic for discussion in this section and thus we will return to this problem shortly.) Now, since $D \equiv$ dose size $= 1$ and so $y = \phi$ (see (16.10)), we calculate the transfer function of the system (A, B, C) as the Laplace transformation

$$(23.10) \qquad \hat{\phi}(s) = \hat{y}(s) = \hat{x}_1(s) = A_1/(s - \lambda_1) + A_2/(s - \lambda_2)$$

$$\equiv (s + c_2)/(s^2 + c_1 s + c_3)$$

where the vector of transfer function coefficients is

$$(23.11) \qquad c \equiv \begin{bmatrix} c_1 \\ c_2 \\ c_3 \end{bmatrix} = \begin{bmatrix} -\lambda_1 - \lambda_2 \\ -A_1\lambda_2 - A_2\lambda_1 \\ \lambda_1\lambda_2 \end{bmatrix}.$$

Upon substituting (23.9) into (23.11), we get

$$(23.12) \qquad c = (2.509, \ 0.8178, \ 0.8419)^T.$$

On the other hand, if the transfer function $\hat{\phi}(s)$ is computed from (17.2), there obtains (see (16.11))

$$\hat{\phi}(s) = (s + \gamma_1)/(s^2 + \gamma_2 s + \gamma_3)$$

where

$$\gamma_1 = a_{12} \equiv \alpha_3$$

$$(23.13) \qquad \gamma_2 = a_{21} + a_{01} + a_{12} \equiv \alpha_1 + \alpha_2 + \alpha_3$$

$$\gamma_3 = a_{01}a_{12} \equiv \alpha_2\alpha_3 .$$

Comparing (23.10) and (23.13) gives the nonlinear system

$$\alpha_1 + \alpha_2 + \alpha_3 = c_1$$

$$(23.14) \qquad \alpha_3 = c_2$$

$$\alpha_2\alpha_3 = c_3 ,$$

a system of type (22.8), which we studied in detail in the last section. The necessary conditions of Theorem 22.4 are satisfied for the c-vector in (23.12), and this system $F(\alpha) = 0$ is easily solved (as in Theorem 22.5) to get

$$\alpha^* = \begin{bmatrix} \alpha_1^* \\ \alpha_2^* \\ \alpha_3^* \end{bmatrix} = \begin{bmatrix} a_{21} \\ a_{01} \\ a_{12} \end{bmatrix} = \begin{bmatrix} 0.6617 \\ 1.0295 \\ 0.8178 \end{bmatrix}.$$

Hence the system compartmental matrix is (uniquely) identifiable as

$$
A = \begin{bmatrix} -1.6912 & 0.8178 \\ 0.6617 & -0.8178 \end{bmatrix} .
$$

Once we know the linear model (16.5) for the tracer kinetics, with A as identified here, and assuming as usual that the tracee behaves approximately the same, we now know the (first order) dynamics of the system.

Exercise 23.1.   We can check the values a_{21}, a_{01}, and a_{12} by other methods.

(a)   Use the first Markov parameter $M_1 = CAB$ (16.20) to compute a_{21}.

(b)   Recall Hamilton's formula (16.23) to compute

$$
a_{01} = 1/ \int_0^\infty x_1(t)dt = -1/ \left(\frac{A_1}{\lambda_1} + \frac{A_2}{\lambda_2} \right) .
$$

(c)   Use the symmetric function equation $s_1 = -\sigma_1$ (16.19) to find a_{12}.

Once the compartmental matrix A has been identified, it is always advisable to run a computer simulation wherein the differential equation model $\dot{x}(t) = Ax(t)$ along with appropriate initial conditions is solved numerically at some of the times for which data are available. For instance, in the above two-compartment example, the differential equation system

$$
\dot{x}_1 = -1.691\ x_1 + 0.818\ x_2
$$

$$
\dot{x}_2 = 0.662\ x_1 - 0.818\ x_2
$$

along with $x_1(0) = 1$, $x_2(0) = 0$, was solved numerically by DVERK (a Runge-Kutta scheme) in the IMSL package [131] at the points t = 0.5, 1.0, 1.5, ..., 10., to see if the data of Table 23.1 is essentially reconstructed:

```
PROGRAM CHECK
INTEGER N, IND, NW, IER, K
REAL X(2), C(24), W(2, 9), T, TOL, TEND
```

```
        EXTERNAL F
        NW = 2
        N = 2
        T = 0.
        X(1) = 1.
        X(2) = 0.
        TOL = 0.0001
        IND = 1
        TEND = 0.5 (successively change)
        CALL DVERK (N, F, T, X, TEND, TOL, IND, C, NW, W, IER)
        PRINT 9, X(1)
9       FORMAT (F 8.4)
        STOP
        END


        SUBROUTINE F(N, T, X, XP)
        INTEGER N
        REAL X(N), XP(N), T
        XP(1) = - 1.692 * X(1) + 0.818 * X(2)
        XP(2) =   0.663 * X(1) - 0.818 * X(2)
        RETURN
        END
```

A sample of the output of CHECK is:

TEND	X_1(TEND)
0.5	0.4630
1.0	0.2560
5.0	0.0334
8.0	0.0101

23C. Nonlinear Least-Squares

Let us now return to the core problem of fitting the general sum of exponentials (23.6) to data of decay measurements. The goal of course is to determine the
"amplitudes" A_i and "decay constants" λ_i. The usual method is to set up a nonnegative objective function J defined on some domain Δ in real parameter space whose
elements satisfy any constraints of the system. With data observation vectors d_1,
d_2, ..., d_m, at discrete times $0 < t_1 < t_2 < \ldots < t_m$, we will define

$$J(A_1, \lambda_1, \ldots, A_n, \lambda_n) \equiv \sum_{k=1}^{m} \| y(t_k) - d_k \|_2^2$$

and then solve the nonlinear least-squares problem if minimizing J over the 2n-
dimensional parameter domain Δ. The functional J measures the closeness of fit of
the model, where we say that the parameter vector $p^{(1)}$ achieves a closer fit than
$p^{(2)}$ provided $J(p^{(1)}) < J(p^{(2)})$. Whether or not there exists a unique global minimum p^* of $J(p)$ on the whole of Δ is not generally known.

Our standard approach to this problem is to use a robust minimizer such as
ZXSSQ or ZXMIN from a reputable source such as the IMSL package. To illustrate in
the case of the data of Table 23.1, let the second column be d_1, ..., d_{13} (m = 13)
corresponding to the first column times t_1, ..., t_{13}. Then we look for a global
minimum on an appropriate Δ of the functional

$$J(A_1, \lambda_1, A_2, \lambda_2) \equiv \sum_{k=1}^{13} [y(t_k) - d_k]^2$$

in which we use the y function of (23.8). A code, using ZXMIN (a quasi-Newton
method) is listed below:

```
PROGRAM BLE
EXTERNAL FUNCT
DIMENSION X(4), H(10), G(4), W(12)
N = 4
NSIG = 4
MAXFN = 1000
```

```
      IOPT = 0

      DATA X/.91, - 1.8, .09, - .23/

      CALL ZXMIN (FUNCT, N, NSIG, MAXFN, IOPT, X, H, G, F, W, IER)

      PRINT 9, X, F

9     FORMAT (F 18.6)

      STOP

      END


      SUBROUTINE FUNCT (N, X, F)

      DIMENSION X(N), T(13), D(13), FA(13), EX(13)

      DATA T/0., .5, 1., 1.5, 2., 3., 4., 5., 6., 7., 8., 9., 10./

      DATA D/1., .461, .259, .17, .121, .0722, .0451, .0319, .024, .0182, .0141,
               .01, .0094/

      DO 5  I = 1, 13

      EX(I) = X(1) * EXP(X(2) * T(I)) + X(3) * EXP(X(4) * T(I))

      FA(I) = (EX(I) - D(I)) ** 2

5     CONTINUE

      SUM = 0.

      DO  6  J = 1, 13

      SUM = SUM + FA(J)

6     CONTINUE

      F = SUM

      RETURN

      END
```

In this program, $X(1) = A_1$, $X(2) = \lambda_1$, $X(3) = A_2$ and $X(4) = \lambda_2$. The DATA X line in the above listing furnishes initial guesses of the parameters:

$$A_1^0 = 0.91 \qquad\qquad A_2^0 = 0.09$$

(23.15)

$$\lambda_1^0 = - 1.80 \qquad\qquad \lambda_2^0 = - 0.23 .$$

(We will discuss arriving at these values momentarily.) The output of program BLE is displayed in (23.9). As a simple check, we note that $x_1(0) = A_1^* + A_2^* = 0.9993 \doteq 1$, as it should be in the model (see (23.8)) since we have a unit input applied at time zero. To be secure against the possibility that the parameter values (23.9) provide only a local minimum of J, some random starting guesses were substituted for (23.15), such as

DATA X/.5, - .77, .5, - .182/,

DATA X/.3, - 1.4, .7, - .5/.

All of these initial points gave rise to the same final estimates (23.9).

Exercise 23.2. Write and execute a program such as BLE to do the exponential curve-fitting of (23.8) to the data of Table 23.1.

23D. Initial Parameter Estimates

Let us now see how the initial guesses suggested at line (23.15) were determined. The process to be described here is called exponential curve peeling [134, p. 103], [232], and is a suitable method provided the number of terms n in the sum of exponentials is small, say $n \le 3$, and if the decay constants λ_i are reasonably well separated. In the analysis of many experiments using tracers, steady-state conditions are normally in effect (recall equation (7.2)) and are necessary for an accurate estimation of the terminal exponential decay constant of the disappearance of tracer. Towards the end of the time interval over which data are taken, only this exponential term of the curve is remaining (see (12.14)) and consequently the intercept and decay constant of this component of the curve are often estimated first.

For our model (23.8), let us assume that $\lambda_1 < \lambda_2 < 0$. Then for large time t, the contribution from the first exponential term will have virtually died out, and so the terminal portion of the curve y(t) is asymptotic:

$$y(t) \underset{\sim}{} A_2 e^{\lambda_2 t}, \quad t \text{ large.}$$

Thus if we define $z(t) \equiv \ln y(t)$ and $a \equiv \ln A_2$, then for large t, there results $z \sim a + \lambda_2 t$, which is a straight line in (t, z)-coordinates. Does our data support this? Let us plot the points (0, 1), ..., (10, 0.0094) from Table 23.1 on semilog paper and see if the last few points turn out to fall roughly in a straight line. SUch a graph in fact demonstrates that the final four points do lie on a straight line. (That they do aids in validating the model, for if the largest time data points are not linear, then the exponential model y(t) is inappropriate and thus we should rethink the modeling process.) Hence we do a linear least-squares fit of $z = a + \lambda_2 t$ to the points

$$(7, \ln 0.0182), \ (8, \ln 0.0141), \ \dots, \ (10, \ln 0.0094),$$

to arrive at $a = -2.41$ and $\lambda_2 = \cdot 0.23$. Thus $A_2 = \exp(a) = 0.09$. Now since

$$y(t) - A_2 e^{\lambda_2 t} = A_1 e^{\lambda_1 t}$$

by (23.8), we should have

$$(23.16) \qquad w(t) \equiv \ln[y(t) - A_2 e^{\lambda_2 t}] = b + \lambda_1 t$$

where $b \equiv \ln A_1$. Compute the quantity within the square brackets [·] of (23.16) for t = 0, .5, 1., ..., 6., using the data of Table 23.1 substituted for y(0), y(.5), y(1.), ..., y(6.), and also utilizing the values $A_2 = 0.09$ and $\lambda_2 = -0.23$. The resulting numbers plotted on semilog paper should also follow a straight line, according to (23.16). Upon carrying out a linear least-square fit of $w = b + \lambda_2 t$ to this data, we get the estimates $b = -0.094$ and $\lambda_1 = -1.8$. Hence $A_1 = \exp(b) = 0.91$. In summary then, our desired four estimates are the values given in (23.15). (Jacquez [134, p. 104] presents a 3-term exponential curve peeling example.)

Exercise 23.3.   Carry out the details of this curve peeling process, as explained above, for the data of Table 23.1.

There are clearly limitations curtailing the usefulness of this peeling technique. The method can be used regardless of the number of exponential terms making up $y(t)$. However, in practice, each exponential term requires a number of data points to determine its amplitude and decay constant, and since these data points are then discarded in the computation of the remaining terms, there is usually a limitation to the number of exponential terms that can be detected and estimated since such a sizable data base is seldom available. (For computerized studies, see [156], [69].)

Initial estimates for A_1, λ_1, A_2, and λ_2 can be derived by other means. One such method is to split the data set $\{(t_k, d_k)\}$ into two groups where we (1) fit $y = A_1 \exp(\lambda_1 t)$ by ln-transformation to the data $\{(t_k, d_k)\}$, $k = 1, 3, 5, \ldots$, to get estimates for A_1, λ_1; (2) fit $y = A_2 \exp(\lambda_2 t)$ in the same manner to $\{(t_k, d_k)\}$, $k = 2, 4, 6, \ldots$, to compute approximations for A_2, λ_2. Then take $(A_1/2, \lambda_1, A_2/2, \lambda_2)$ as the initial guess [133].

Another approach to getting initial estimates is as follows:

(a) Make an educated guess for the decay constants: λ_1^0, λ_2^0, $\ldots$, λ_n^0 (this may be done by one of the previous methods);

(b) Do linear regression to solve

$$\min_{A_1,\ldots,A_n} J(A_1, \lambda_1^0, A_2, \lambda_2^0, \ldots, A_n, \lambda_n^0)$$

and let the solution A_1^0, $\ldots$, A_n^0 complete the set of initial estimates.

This method is based on the fact that parameters A_i only enter (23.6) linearly and hence can be handled by linear regression once the nonlinear parameters λ_i are determined. Variations based on linearity are also utilized in [148], [133].

Exercise 23.4. With $\lambda_1^0 \equiv -1.8$ and $\lambda_2^0 \equiv -0.23$, use any of the concepts suggested above to find amplitude initial estimates A_1^0 and A_2^0. How do they compare with those of (23.15)? Are they closer to the estimates of the amplitudes shown in (23.9) than those of (23.15)?

23E. Method of Moments

In computing optimal amplitude and decay constant estimates for sums of exponentials such as in (23.9), we have used standard nonlinear optimization algorithms such as ZXMIN (quasi-Newton scheme) or ZXSSQ (Levenberg-Marquardt, Gauss-Newton schemes) in the IMSL package. Both of these codes are set up for general nonlinear minimization problems and were not explicitly designed for sums of exponentials. Methods designed just for parameter estimation in sums of exponentials have been developed, with varying degrees of success. It appears that in most of these cases the methods developed have turned out to yield estimates which are not very accurate and as a consequence should probably be thought of as initial estimates to start up some nonlinear minimizer (as discussed above) to refine them. These methods (other than the nonlinear least-squares) fall under the general categories of method of moments, and Fourier transformation techniques. We will discuss here only a method of moments approach to the problem.

The "method of moments" type of computational procedure serves up some interesting analysis. The scheme to be presented here is for identifying A_i, λ_i (and possibly the number n of components) in (23.6) from discrete real data $(t_1, d_1), \ldots,$ (t_m, d_m) for $y(t)$, and is primarily based on results in [94], [132]. We start by assuming that $\lambda_1, \lambda_2, \ldots, \lambda_n$ are distinct nonzero real numbers and that the A_i are all nonzero. For some prescribed interval [a, b] (containing the time values at which the discrete data are taken), define X to be the n-dimensional vector space spanned by the linearly independent functions

$$\phi_i(t) \equiv \exp(\lambda_i t) \qquad i = 1, 2, \ldots, n.$$

The method of moments algorithm is based on a set of 2n linear functionals h_1, h_2, $\ldots$, h_{2n}, each mapping X into the reals R, having the properties (a) for each $x \in X$ and its time derivative $\dot{x}$, we have

$$(23.17) \qquad h_i(\dot{x}) = h_{i+1}(x) \qquad i = 1, 2, \ldots, 2n-1;$$

(b) the h_i are linearly independent. In particular, we have in mind the functionals (which arise naturally from consideration of the concept of integral moments)

$$(23.18) \qquad h_i(x) \equiv (-1)^{i-1} \int_a^b x(t) w^{(i-1)}(t) dt \qquad i = 1, \ldots, 2n$$

where $w(t)$ is a preselected known weighting function with sufficient continuous derivatives on $[a, b]$, and also satisfying the end point conditions

$$w^{(i-1)}(a) = w^{(i-1)}(b) = 0 \qquad i = 1, \ldots, 2n-1.$$

A simple example of such a weight function is $w(t) = [(t-a)(t-b)]^{2n-1}$. The choice of w allows flexibility in weighting the data base.

Exercise 23.5. For any $x \in X$, show that the functionals h_i defined in (23.18) satisfy requirement (23.17).

Now for $y(t)$ in (23.6), each derivative $y^{(k)}$, $k = 0, 1, 2, \ldots$, is in X.

Exercise 23.6. Show that $y, \dot{y}, \ldots, y^{(n-1)}$ form a linearly independent set in X, while $y, \dot{y}, \ldots, y^{(n)}$ form a linearly dependent set in X.

Because of this exercise, we can say that there exist constants $c_0, c_1, \ldots, c_{n-1}$, not all zero, such that

$$(23.19) \qquad y^{(n)} + c_{n-1}y^{(n-1)} + \ldots + c_0 y \equiv 0$$

on $[a,b]$. Moreover, the λ_i are roots of the associated characteristic equation

$$(23.20) \qquad p(\lambda) \equiv \lambda^n + c_{n-1}\lambda^{n-1} + \ldots + c_0 = 0.$$

Our first step is in the identification of the unknown polynomial coefficients c_i. Operating with the linear functional h_i on (23.19) gives the system of equations $(i = 1, \ldots, n)$

$$c_0 h_i(y) + c_1 h_i(\dot{y}) + \ldots + c_{n-1}h_i(y^{(n-1)}) = - h_i(y^{(n)}).$$

We can place this system into matrix-vector form $Gc = r$, where

$$c \equiv (c_0, c_1, \ldots, c_{n-1})^T,$$

$$r \equiv - (h_1(y^{(n)}), h_2(y^{(n)}), \ldots, h_n(y^{(n)}))^T,$$

$$G \equiv [g_{ij}]_{n \times n}, \qquad g_{ij} \equiv h_i(y^{(j-1)}).$$

Exercise 23.7. Prove that we can express g_{ij} as $h_{i+j-1}(y)$, thus showing that the G matrix is symmetric. Also verify that the ith component of r is $r_i = -h_{i+n}(y)$. These two equalities show that both G and r may be computed directly from $y(t)$ (by numerical integration using the data $d_1, \ldots, d_m$ for y).

Exercise 23.8. Prove that G is nonsingular. [Hint: Let $\{z_1, \ldots, z_n\}$ be a basis for X and let $\{g_1, \ldots, g_n\}$ be its dual basis; use the condition $g_j(z_i) = \delta_{ij}$.]

Since G is invertible, then a good symmetric linear equation solver (such as LEQ1S in the IMSL package) can be used to solve Gc = r for c.

Once c is known, then (23.20) can be solved for roots $\lambda_1, \ldots, \lambda_n$. Hence only the amplitudes $A_1, \ldots, A_n$ are yet to be found. Now the amplitudes enter the problem only linearly and thus they may be determined by multilinear regression analysis, as discussed before. However, since the $h_i(y)$ have already been deter-mined by numerical integration and are required in the computation of G and r, another avenue is open for us. Apply h_i to each side of (23.6) to get

$$(23.21) \qquad \sum_{j=1}^{n} h_i(\phi_j)A_j = h_i(y) \qquad i = 1, 2, \ldots, n.$$

If we introduce

$$u \equiv (A_1, A_2, \ldots, A_n)^T,$$

$$Q \equiv [q_{ij}]_{n \times n} \equiv [h_i(\phi_j)],$$

$$s \equiv (h_1(y), h_2(y), \ldots, h_n(y))^T,$$

then (23.21) is the n-square linear system Qu = s. As noted before, the s-vector can be computed from the data for y, and the entries of Q can be calculated by direct integration since λ_j are now known and so the functions $\phi_j(t) = \exp(\lambda_j t)$ are known. Moreover, since

$$q_{1+1,j} = h_{i+1}(\phi_j) = h_i(\dot{\phi}_j) = h_i(\lambda_j \phi_j) = \lambda_j h_i(\phi_j) = \lambda_j q_{ij},$$

we can generate the entire j^{th} column of Q (j = 1, ..., n) by the recursion formula

$$q_{kj} = \lambda_j^{k-1} q_{1j} \qquad\qquad k = 2, 3, \ldots, n$$

and hence the matrix Q is expressible as Q = VD where V is the <u>Vandermonde matrix</u>

$$V \equiv \begin{bmatrix} 1 & 1 & \cdots & 1 \\ \lambda_1 & \lambda_2 & \cdots & \lambda_n \\ \cdots & \cdots & \cdots & \cdots \\ \lambda_1^{n-1} & \lambda_2^{n-1} & \cdots & \lambda_n^{n-1} \end{bmatrix}$$

and D is the matrix diag $\{q_{11}, q_{12}, \ldots, q_{1n}\}$. The equation Qu = s becomes VDu = s. Let $v \equiv Du$. Then we are dealing with the linear system Vv = s. Here the Vandermonde matrix V is nonsingular since the λ_i's are distinct. Thus we can solve for v. If we are careful in our choice of weights, then $h_1(\phi_i) = q_{1i}$ will be nonzero for all i, and so D is invertible. Then $u = D^{-1}v$ so that the ith amplitude is v_i/q_{1i}, or

$$(23.22) \qquad A_i = v_i/h_1(\phi_i) \qquad\qquad i = 1, 2, \ldots, n.$$

<u>Example [94]</u>. The two component function y(t) in (23.8) with

$$(23.23) \qquad \begin{array}{ll} A_1 = 1 & A_2 = 1 \\[2mm] \lambda_1 = -1/10 & \lambda_2 = -1/30 \end{array}$$

was used to generate the data values y(0), y(1), y(2), ..., y(20). A curve-fitting of these 21 points by the previously described method is done using the weight function $w(t) = [\sin(\pi t/20)]^{2n-1}$, where n = 2, and Simpson's Rule for the numerical integration, to get the results

$$(23.24) \qquad \begin{array}{ll} A_1^* = 1.047 & A_2^* = 0.957 \\[2mm] \lambda_1^* = -0.098 & \lambda_2^* = -0.032 \end{array}$$

which reconstructs the function quite well. If more data points are used, then an
even better fit can be produced.

There are difficulties that can arise with the above method:

(a) The roots λ_1, λ_2, ..., λ_n of $p(\lambda) = 0$ may not turn out to be real and/or
distinct;

(b) Solving $Gc = r$ and/or $Qu = s$ may be ill-conditioned;

(c) Considerable data is needed (and not often available) to provide good numerical
estimates of the functionals.

On the other hand, some advantages of the method are that no starting values are
needed, and the computer CPU time can compare favorably with the more standard
method of nonlinear least-squares.

Other methods closely akin to the method of moments are available in references [99], [179], and can be used for finding initial estimates of the parameters
A_i and λ_i which then may be used in a more refined technique such as nonlinear
least-squares.

23F. Other Methods of Parameter Estimation

The other category of methods mentioned earlier was that of Fourier transformation techniques. These methods appear to have started with Gardner et al.
[103] and been modified in [102], [184], [185], [118], [50], [181], [219]. Discrete
Fourier transformations are also used in deconvolution methods for determining ϕ
from known u and measured y in (23.3). See [134, p. 114].

We should mention briefly some nonlinear minimization programs especially
designed with sums of exponentials in mind. One such numerical technique goes back
to the work of Worsley and Lax [243] involving the principle of least-squares,
followed by Berman's efforts [30], [33], and Davis and Ottaway [74]. More recently
algorithms have been studied for finding the optimal least-squares exponential sum
approximations to sampled data subject to the additional constraint that the
amplitudes A_i are positive [95] (also see [139]). The numerical scheme in [95] uses
the exponential of a triangular matrix [165] to obtain divided differences of
exponentials which overcomes certain problems in the natural ill-conditioning of

the fitting of sums of exponentials. The principle of least-squares can also be classified under likelihood methods, and such curve-fitting techniques are presented in [158], [141], and [164].

23G. Positive Amplitudes

The _positive coefficient_ aspect of this theory is interesting and is suitable for certain problems (although there are many biological situations where the amplitudes A_i in (23.6) are not all positive). We might recollect that in our discussion of the curve-peeling process for the data of Table 23.1, we defined real parameters $a \equiv \ln A_2$ and $b \equiv \ln A_1$, thereby tacitly assuming that A_1 and A_2 are positive. We can put substance behind this assumption. Let us reconsider equations such as (20.12) or (20.19) involving the entries $z_{ij}^{(k)}$ of the _component matrices_ Z_k of the compartmental matrix A. The "amplitudes" in these sums of exponentials were in fact the $z_{ii}^{(k)}$, which are positive in the curve-fit given in (20.17). That these elements are positive is not a fluke, for we have at our disposal Theorem 16.1 which gives conditions on the matrices A, B, and C of our system (23.1) for which we can expect an output function with positive amplitudes. It is most useful to know in advance that the coefficient parameters must be positive, for in such cases a particular sums of exponentials algorithm such as that in [95] can be utilized. Moreover, the sums of exponentials form (23.6) is not necessarily realizable by a compartmental model; however, a sufficient condition for realizability is that $A_i > 0$ for all $i = 1, 2, \ldots, n$ [123], [47].

Exercise 23.9. Let A be any $n \times n$ real matrix (not necessarily compartmental) with distinct eigenvalues, and symmetrizable by a positive definite diagonal matrix. Show that any diagonal entry of any component matrix of A is nonnegative.

Exercise 23.10. Let A satisfy the conditions of the previous exercise and in addition have eigenvalues $\lambda_1 < \lambda_2 < \ldots < \lambda_n \leq 0$. Consider the model $\dot{x}(t) = Ax(t)$, $t \geq 0$, with $x(0) = De_i$ for constant $D > 0$, e_i the ith column of the $n \times n$ identity matrix, and fixed index i. Show that $x_i(t)$, the ith entry in $x(t)$, is now ensured to be monotone decreasing (also see [182], [119]).

An observation coming off of Exercise 23.9 is that if A is any real matrix satisfying the stated hypothesis, then inequality

(23.25)
$$\max_{1\leq i\leq n} |a_{ii}| \leq |\lambda_1|$$

holds, which was discussed back in Section 20 (see (20.8)). Another observation that can be deduced from this work is that the "sensitivity"

$$\partial\lambda_k/\partial a_{ii} = z_{ii}^{(k)} \geq 0 \qquad i, k = 1, 2, \ldots, n,$$

of the kth eigenvalue of A to small changes just in that model parameter a_{ii} is nonnegative (see (20.25), Theorem 20.1, and [5]).

23H. Curve-Fitting Sums of Exponentials is Ill-Posed

Some final comments on curve-fitting sums of exponentials are in order. Numerous scientific problems require fitting sums of exponentials (23.6) to experimental data for y to determine A_i and λ_i. This problem turns out to be difficult because it is "ill-posed". By ill-posed, we mean that if perfect data is generated by a preselected sum of exponentials, say three, and this data is perturbed slightly, then a sum of two exponentials can be found, or a different three exponential sum, which fits (in a least-squares sense) the perturbed data as well as the original three exponential sum [144, pp. 278-279]. This basic inherent difficulty of fitting exponentials to data cannot be remedied by any modified mathematical procedure since the difficulty lies with the ill-conditioning of the problem, that is, with the extreme sensitivity of the amplitudes and decay constants to very small changes of the data. The sole remedy is to increase both the amount and accuracy of the data by improved and new experimental methods.

23I. Fitting the Differential Equation Model Directly to Data

Let us turn now to a completely different approach to the parameter estimation problem of matching system (23.1) to data so as to determine the parameter vector α. In this new approach we fit the differential equation model itself directly to the data to estimate α instead of first forming and fitting solution (23.6) to data,

and then estimating α from the parameters of the solution function through identification equations such as (23.14) or (22.8). (If the model had been nonlinear,
then in all probability no closed form solution for the model could be written, and
so a curve-fitting of the solution form to data would be impossible; hence in such
a case an approach of the current type must be taken.) The actual fitting of the
model directly to data is usually handled by solving the differential equations
numerically in conjunction with a nonlinear least-squares minimization. In fitting
the linear rate equations (23.1) directly to the data set, we have the advantage of
avoiding any assumptions to be made about the multiplicity of the eigenvalues of the
system matrix A, and of avoiding being forced to form the nonlinear equations (22.8)
and solve them.

   To carry out such an approach on model (23.1), we recast the estimation problem slightly from what it was before. Again, the usual process is to construct a
nonnegative objective functional $J(\alpha)$ defined on an appropriate domain Ω in the real
parameter space R^{ν}. Given the data vectors d_1, d_2, ..., d_m, at discrete times
$0 < t_1 < t_2 < ... < t_m$, we wish to select α so as to minimize over all of Ω the sum
of squares functional

$$(23.26) \qquad J(\alpha) \equiv \sum_{k=1}^{m} \| y(t_k, \alpha) - d_k \|_2^2 .$$

The difference between this problem and the previous sum of squares approach is that
at the jth step of this least-squares process, the most recent parameter estimate
$\alpha^{(j)}$ is fed into the differential equations which are integrated numerically to
provide $x(t, \alpha^{(j)})$ and thus $y(t, \alpha^{(j)}) = C(\alpha^{(j)}) x(t, \alpha^{(j)})$, which is then used in
(23.26). Typically, data are obtainable from only one or a few compartments. Hence
it is quite possible that one may not have data on enough compartments to determine
all the parameters (e.g., if $\nu > pm$, where p is the row dimension of C in (23.1)).
Even if there are adequate data, the fact that only a few components of x are
sampled can cause great hardship for any parameter estimation method, but the above
approach is as reasonable an approach to the problem as seems possible to take.

<u>Example</u>. An example serves best to illustrate this method. Let us suppose our model is that of (23.7),

$$\dot{x}_1 = (- a_{21} - a_{01})x_1 + a_{12}x_2$$

(23.27)

$$\dot{x}_2 = a_{21}x_1 - a_{12}x_2$$

for all $t \geq 0$, with $x_1(0) = 1$, $x_2(0) = 0$, $C = [1 \ 0]$, and that we are going to fit this model to the lipoprotein data of Table 23.1.

Since there are $\nu = 3$ fractional transfer coefficients $\alpha_1 \equiv a_{12}$, $\alpha_2 \equiv a_{21}$, and $\alpha_3 \equiv a_{01}$ to estimate, the problem is a three parameter fitting problem of minimizing $J(\alpha)$ over the positive orthant

$$\Omega \equiv \{\alpha \equiv (\alpha_1, \alpha_2, \alpha_3) \in R^3 : \alpha_i > 0\} .$$

Suppose we make direct use of the data and our theory of compartmental models to reduce the number of parameters to be estimated by the least-squares process. We can do this for the present model since it is single exit [4], [8], and we have available formula (16.23) (also see Exercise 23.1) to compute the parameter

$$\alpha_3 = a_{01} = 1/ \int_0^\infty x_1(t)dt = 1/ \int_0^\infty y(t)dt$$

by numerical integration of the data for output y. Let us assume this has been done with the resulting value $\alpha_3 = a_{01} = 1.029$.

We now proceed by making initial estimates for the remaining parameters α_1 and α_2. This may be done by talking to some of the experts in the field from which the problem is taken, and/or estimating the derivatives $\dot{x}$ by some finite difference approximation (most likely wanting to first "smooth" the data points and then use the smoothed points for estimating the derivatives) and then utilizing what data there is for x to set up linear equations or a linear least-squares problem to get the initial guesses α_1^0, α_2^0. Once these starting values are picked for α_1 and α_2, the system (23.27) is integrated numerically to yield a projected data set for comparison with the observed data d_1, ..., d_m of Table 23.1. Then the least-squares functional (23.26) is used to minimize the differences between the calculated and experimental data set. The new parameter estimates thus produced are given to the differential equations which are again integrated numerically to provide a new computed data set. The following program carries out this process for our example (23.27) with $y(t) = x_1(t)$ data in Table 23.1:

```
      PROGRAM DIRFIT
      DIMENSION X(2), F(12), XJAC(12, 2), XJTJ(3), WORK(37), PARM(4)
      EXTERNAL FUNC
      IXJAC = 12
      M = 12
      N = 2
      NSIG = 4
      EPS = .0001
      DELTA = .0001
      MAXFN = 500
      IOPT = 1
      DATA X/.8, .6/
      CALL ZXSSQ(FUNC, M, N, NSIG, EPS, DELTA, MAXFN, IOPT, PARM, X,
     # SSQ, F, XJAC, IXJAC, XJTJ, WORK, INFER, IER)
      PRINT 9, X, SSQ
    9 FORMAT (F14.8)
      STOP
      END

      SUBROUTINE FUNC(X, M, N, F)
      DIMENSION X(N), F(M), XO(2), C(24), XP(2), T(12), D(12), W(2,9)
      COMMON Q, R
      EXTERNAL FU
      Q = X(1)
      R = X(2)
```

```
        T1 = 0.

        NW = 2

        N1 = 2

        X0(1) = 1.

        X0(2) = 0.

        TOL = .0001

        IND = 1

        DATA T/.5, 1., 1.5, 2., 3., 4., 5., 6., 7., 8., 9., 10./

        DATA D/.461, .259, .17, .121, .0722, .0451, .0319, .024,

        # .0182, .0141, .01, .0094/

        DO 2 IT = 1, 12

        TEND = T(IT)

        CALL DVERK (N1, FU, T1, X0, TEND, TOL, IND, C, NW, W, IERR)

        F(IT) = X0(1) - D(IT)

2       CONTINUE

        RETURN

        END


        SUBROUTINE FU(N1, T1, X0, XP)

        INTEGER N1

        REAL X0(N1), XP(N1), T1

        COMMON Q, R

        XP(1) = (-R - 1.029) * X0(1) + Q * X0(2)

        XP(2) = R * X0(1) - Q * X0(2)

        RETURN

        END
```

This program uses subroutine ZXSSQ (modified Levenberg-Marquardt for sums of squares) and DVERK (Runge-Kutta-Verner fifth and sixth order differential equation solver) from the IMSL package. In the program there are $N = 2$ parameters: $\alpha_1 \equiv Q \equiv x(1)$ and $\alpha_2 \equiv R \equiv x(2)$. The number of data points is $M = 12$, where the initial point

(0, 1) appears in the form of the initial condition Tl = 0., XO(1) = 1. (along with

XO(2) = 0.). The function F(IT) = XO(1) - D(IT), for each value of IT, is the dif-

ference inside the norm of (23.26). With a variety of initial guesses such as

    DATA X/0.8, 0.6/

    DATA X/0.5, 0.9/

    DATA X/1.2, 0.2/

the output of program DIRFIT is in all cases $\alpha_1^* = 0.82$ and $\alpha_2^* = 0.66$ with a sum of

squares $J(\alpha^*) = 1.2 \times 10^{-4}$ at the final parameter values. This matches the output

we received from (23.14), which was derived from fitting the sums of exponentials

solution form to the data of Table 23.1.

Exercise 23.11.   Consider the (nonlinear) logistic model

$$\dot{x}(t) = kx(t) - cx^2(t), \qquad t \geq 0,$$

$$x(0) = 1.$$

We wish to fit this model to the (admittedly rather skimpy) data set for x(t)

t_i	1	2	3
d_i	1.220	1.488	1.815

to determine the positive parameters k and c. Approximate $\dot{x}(t)$ by the finite dif-

ference x(t + 1) - x(t) (or by a central difference) to get crude estimates of

$k^0 = 0.22$ and $c^0 = 0.0002$. Write and execute a program in the flavor of DIRFIT to

get refined least-squares estimates of k and c.

23J. Modulating Function Method

As the final technique in this section on parameter estimation, we discuss a

"modulating function method" [7] for estimating unknown parameters which appear in

either linear or nonlinear ordinary differential equation models as multipliers of the dependent variables (as in our tracer equations). The concept (in theory) works for systems of differential equations whose solution x is known (at least at discrete times) for some of the components x_i, and may be unknown in the other components. By numerical quadrature, a set of linear equations in the parameters (and other quantities) is developed which can be either directly solved or handled by linear least-squares to obtain estimates of the parameters of the model. Simultaneously, estimates of the unknown components are also obtained. The method is similar in flavor to the method of moments presented above (and has many of the same drawbacks).

Let us assume that the equations of the differential equation system are so ordered that the first p components x_i are known (experimentally observed) at the common times $a \equiv t_0 < t_1 < \ldots < t_m \equiv b$. We then choose a finite collection of functions, $\{\phi_k\}$, to be called "modulating functions," from the space $C^1[a, b]$, say (sufficient smoothness conditions are needed so that integration by parts is allowable), where the space is equipped with the usual inner product

$$<f, g> \equiv \int_a^b f(t)g(t)dt.$$

We form the inner product of each ϕ_k with the components of the particular differential equation model being used,

$$\dot{x}_i(t) = F_i(t, x(t), \alpha), \qquad a \leq t \leq b, \qquad i = 1, 2, \ldots, n,$$

$$x(a) = x^0$$

to get

(23.28) $<\phi_k, \dot{x}_i> = <\phi_k, F_i>$

and then integrate the left-hand side of (23.28) by parts:

(23.29) $\phi_k(b)x_i(b) - \phi_k(a)x_i(a) - <\dot{\phi}_k, x_i> = <\phi_k, F_i>.$

The integration by parts removes $\dot{x}$ and restates the equations just in terms of x (on which we have some information). In (23.29) we replace the entries in the first p components of x by the elements in d_j, j = 1, 2, ..., m. This allows the inner products in (23.29) which involve observed components to be calculated by numerical quadrature via the quadrature formula

$$(23.30) \quad \int_a^b \psi(t)dt \approx \sum_{j=0}^m c_j \psi(t_j)$$

where the c_j's are specified according to the numerical integration technique being utilized. Each equation in (23.29) will not only contain unknown parameters in α, but also unknown combinations involving the unsampled components at the observation times:

$$x_i(t_j) ; \quad i = p + 1, p + 2, ..., n; \quad j = 1, 2, ..., m.$$

Order all the unknowns (including the entries of α) into a larger parameter vector β. By selecting a sufficiently large number of modulating functions, equations (23.29) for an algebraic system (or an overdetermined system) that can be solved for β. This system is linear provided the differential equations are of a certain admissible type (as we will see below).

Such is exactly the case in the linear differential equation model (23.1) of interest in this section. Let us take a concrete example of (23.1) for ease of understanding. Suppose in (23.1) the number of equations is n = 2 and the equations are those of model (23.7),

$$\dot{x}_1 = a_{11}x_1 + a_{12}x_2$$

$$\dot{x}_2 = a_{21}x_1 + a_{22}x_2$$

with initial conditions $x_1(0) = 1$, $x_2(0) = 0$ given. Moreover, we have $a_{11} = -a_{21} - a_{01}$ and $a_{22} = -a_{12}$. Here we take as our three unknown parameters $\alpha_1 \equiv a_{11}$, $\alpha_2 \equiv a_{12}$, and $\alpha_3 \equiv a_{21}$, and discuss fitting the model to the lipoprotein data of Table 23.1 (so $0 \le t \le 10$).

To estimate the α_i s, we proceed through the steps outlined above. Equation (23.29), for each modulating function ϕ_k, becomes

$$\phi_k(10)x_1(10) - \phi_k(0)x_1(0) - \int_0^{10} \dot{\phi}_k(t)x_1(t)dt$$

$$= \alpha_1 \int_0^{10} \phi_k(t)x_1(t)dt + \alpha_2 \int_0^{10} \phi_k(t)x_2(t)dt$$

when i = 1, and for i = 2:

$$\phi_k(10)x_2(10) - \phi_k(0)x_2(0) - \int_0^{10} \dot{\phi}_k(t)x_2(t)dt$$

$$= \alpha_3 \int_0^{10} \phi_k(t)x_1(t)dt - \alpha_2 \int_0^{10} \phi_k(t)x_2(t)dt.$$

By utilizing the initial conditions and quadrature formula (23.30), these last two equations translate into

(23.31)

$$\phi_k(10)x_1(10) - \phi_k(0) - \sum_{j=0}^{m} c_j\dot{\phi}_k(t_j)x_1(t_j)$$

$$= \alpha_1 \sum_{j=0}^{m} c_j\phi_k(t_j)x_1(t_j) + \alpha_2 \sum_{j=0}^{m} c_j\phi_k(t_j)x_2(t_j)$$

and the sister equation

(23.32)

$$\phi_k(10)x_2(10) - \sum_{j=0}^{m} c_j\dot{\phi}_k(t_j)x_2(t_j)$$

$$= \alpha_3 \sum_{j=0}^{m} c_j\phi_k(t_j)x_1(t_j) - \alpha_2 \sum_{j=0}^{m} c_j\phi_k(t_j)x_2(t_j) .$$

In equations (23.31) and (23.32), define $u_j \equiv \alpha_2 x_2(t_j)$ and $v_j \equiv x_2(t_j)$. Thus these two equations contain the vector β consisting of the $2m+2$ unknowns

$$\alpha_1, u_1, u_2, \ldots, u_m, v_1, v_2, \ldots, v_m, \alpha_3 .$$

Notice that all the unknowns of β appear linearly in equations (23.31) and (23.32), which hold for each modulating function ϕ_k. Hence if we pick N different modulating functions, then we can solve these equations for β as a square linear algebraic system (when $N = 2m+2$), or determine β by linear regression (when $N > 2m+2$).

Let us now assume that the parameter vector β has been determined. Hence estimates of α_1 and α_3 are in hand. The unknown parameter α_2 remains to be found. Since α_2 enters as a simple "multiplier" in u_j, and since for each j, $u_j = \alpha_2 v_j$, then for any $v_j \neq 0$, we have $\alpha_2 = u_j/v_j$. Now $u_1, \ldots, u_m$ and $v_1, \ldots, v_m$ have been estimated as a part of determining β. Hence a reasonable choice (assuming all v_1, $\ldots$, v_m are nonzero) for the last parameter is the average

$$\alpha_2 \equiv \left(\sum_{i=1}^{m} u_i/v_i \right)\Big/ m \ .$$

The observations of this last paragraph are really what allow the modulating function method to work, at least as a "linear" problem. Each parameter α_i should always enter the problem as a simple multiple of a term which contains at most one unsampled component of x, and that component, if present, must appear multiplicatively and to the first power in this term. Thus, for instance, the four parameter nonlinear system

$$\dot{x}_1 = \alpha_1 x_2 x_3$$

(23.33)
$$\dot{x}_2 = \alpha_2 x_1 x_2^2 x_3 + \alpha_3 x_3$$

$$\dot{x}_3 = \alpha_4 x_2/(x_1 + t)$$

is admissible by this standard if data for x_1 and x_2 is known (i.e., $p = 2$), but not admissible if data for x_1 alone is available ($p = 1$).

Exercise 23.12.   Verify this last statement for the system of (23.33).

23K. An Antigen - Antibody Reaction Example

As a very simple biological (but nontracer) illustration of the method, let us consider the parameter identification of a bimolecular reaction equation which arises from evidence indicating diffusion control of the reaction of carcinoembryonic antigen (CEA) binding to insoluble anti-CEA. This binding of antigen and antibody can frequently be modeled as a biomolecular reversible reaction [101], and the validity of the model is pursued in [100]. The initial value problem is

$$\dot{x} = k_f(A_0 - x)(B_0 - x) - k_r x, \qquad t \geq 0$$

(23.34)

$$x(0) = 0$$

where A_0 is the initial molar concentration of CEA, B_0 is the initial molar concentration of the immuno-globulin fraction of rabbit antisera to CEA (RaCEA) sites capable of binding CEA, x is the molar concentration of CEA bound to RaCEA as a function of time, and k_f, k_r are the forward and reverse rate constants, respectively. Parameters k_f, k_r, and B_0 are unknown.

In its present form, the differential equation for x does not possess unknown parameters which appear as multipliers in the terms of the model. However, the equation can be rewritten as

$$\dot{x} = \alpha_1(A_0 - x) + \alpha_2(x^2 - A_0 x) + \alpha_3 x$$

where $\alpha_1 \equiv k_f B_0$, $\alpha_2 \equiv k_f$, $\alpha_3 \equiv - k_r$, and these parameters do present themselves in the proper format as multipliers. Moreover, if the α_i s can be estimated by the above technique, then estimates of the desired parameters k_f, k_r, and B_0 are immediately available.

With known parameter $A_0 = 1.0$, the problem now is to identify numerically the α_i based on the experimental observations for x presented in Table 23.2. Selecting

TABLE 23.2.

t	x		t	x
1	0.500		8	0.889
2	0.667		9	0.900
3	0.750		10	0.909
4	0.800		11	0.917
5	0.833		12	0.923
6	0.857		13	0.929
7	0.875		14	0.933

three modulating functions, say $\phi_k(t) \equiv t^{k-1}$, $k = 1, 2, 3$, as elements of $C^1[0, 14]$, equation (23.29) for this model becomes

$$(23.35) \qquad \phi_k(14)x(14) - <\dot{\phi}_k, x> = \sum_{j=1}^{3} \alpha_j <\phi_k, f_j> , \qquad k = 1, 2, 3,$$

where $f_1 \equiv A_0 - x$, $f_2 \equiv x^2 - A_0 x$, $f_3 \equiv x$. Upon computing all the inner products using Simpson's rule and the data, the linear system (23.35) is

$$2.72\alpha_1 - 1.75\alpha_2 + 11.3\alpha_3 = 0.933$$

$$11.3\alpha_1 - 9.52\alpha_2 + 86.7\alpha_3 = 1.780$$

$$86.6\alpha_1 - 77.1\alpha_2 + 82.8\alpha_3 = 9.410 .$$

This system has the solution $\alpha_1 = 0.879$, $\alpha_2 = 0.786$, and $\alpha_3 = -0.00742$, so that the estimates for the model parameters are $B_0 = 1.11$, $k_f = 0.786$, and $k_r = 0.00742$.

Exercise 23.13.   Fill in the details of the calculations for the above example. Refine the estimates for B_0, k_f, and k_r by using these estimates as starting values in a nonlinear least-squares scheme similar to DIRFIT.

If a greater number of modulating functions is chosen so that the linear system is overdetermined and linear least-squares is used, or if an exceptional amount of data are available, values for the parameters B_0, k_f, and k_r can be improved. However, our experience has been that any estimates coming out of this noniterative method should be considered at best as start-up values for some iterative scheme such as a standard nonlinear least-squares minimizer (e.g., DIRFIT) so as to refine them.

23L. Additional Literature on Fitting of Differential Equations to Data

For further information on the direct fitting of differential equations to data, one can see Berman et al. [32]. This reference includes a good compartmental example which could be worked as an exercise. In Patten and Witkamp [180], the rate constants of the differential equation model were determined by fitting models to data with an analog computer. Davis and Ottaway [74] discuss fitting compartmental models to tracer data by conjugate gradient and simplex methods. In particular, they point out difficulties that may be encountered in using the SAAM program [33] for anything other than linear differential equation systems (such as our tracer models) for which it evidently was originally designed. In another variation involving several areas touched on in this section, Silvers et al. [215] fitted a three compartment catenary model to radioisotope decay data, obtaining initial parameter estimates by exponential curve-peeling, which were refined by a minimizer for functions of several variables, and then these refined guesses were used as starting values in the SAAM package. More recently, spline methods have received a lot of attention. Such a method using cubic splines for recovering the $n \times n$ constant system matrix A in $\dot{x} = Ax$ wherein less than the full n components of x are known in tabular form, is presented in [1]. One disadvantage of this proposed method for our problems is that it requires equally spaced data, which we do not normally have in biomedical studies. Also we mention a paper by Varah [233] in which the proposed method consists of first fitting the known data by least-squares using cubic splines (so that the data has in effect been smoothed), and then finding the model parameters by least-squares solution of the differential equation sampled at a set of points. Varah discusses, among his choice of four examples, a case wherein data are only given for some of the components of the differential equation. In this particular case, the lack of information on one of the compoenents is overcome by converting the system of two equations into a single equation of second order involving only that variable for which data measurements are available. This approach works only if the nondata variables can be solved for explicitly, and if they can, then in the case of our linear compartmental model, we are led back to solving nonlinear equations which relate data information to model parameters such as those of (23.14) or, more generally, (22.8). Another method of parameter estimation that has been extensively employed in the biological sciences is quasilinearization [27], [28], [49], [147], [25], [26]. A discussion of the convergence properties of the quasilinearization algorithm for parameter estimation is given by Banks and Groome [18]. In particular, it is found that the algorithm can be only linearly convergent but is quadratically convergent if the sum of squares (23.26)

is exactly zero at the final parameter vector solution. Finally, we note that a
reasonable survey of iterative methods used in biological systems involving some
type of nonlinear least-squares optimization will be found among [25], [26], [56],
[74], [149], [166], [223], [229].

SECTION 24. NUMERICAL SIMULATION OF THE MODEL

24A. Compartmental Model Simulation

To simulate the model $\dot{x} = Ax + b$, $t \geq 0$, $x(0) = x_0$, will, for us, mean to solve the differential equation system numerically. To do this, we must select the input b, initialize the levels in each compartment by choosing x_0, and then determine the model parameters a_{ij} in A to be used. The identification of these fractional transfer coefficients usually comes about in one of two ways. Either data are collected from the experiment in question and numerical methods are applied to estimate the a_{ij} s (see the previous section), or if observations are not possible, then we can talk to some of the experts in the field and put together educated guesses for the values of the a_{ij}. We may only need "reasonable" values for these coefficients if we are not too sure of the validity of the model equations. By trying different sets of transfer parameters $\{a_{ij}\}$, a variety of scenarios can be created and thus the true physical processes can be simulated by the mathematical model. We may wish to test some basic assumption that is present--for example, that the inflow or outflow of material in the system is really proportional to the amount of material in the donor compartment. Partial validation of the model comes from getting a negative answer to the question of does the model lead to any obvious contradictions? The normal checks imposed are:

(a) Do observations match the simulated results?

(b) Consideration of the limiting cases - e.g., when no material is injected; when a very large dose is injected; when $t \to \infty$.

(c) Common sense.

Numerical simulation of our linear model, $\dot{x} = Ax + b$, can of course be done by using the closed form solution (10.5). However, if n is even moderately large, studies [165] have shown that directly computing the solution vector (10.5) for the model, although theoretically possible, may be excessively time-consuming and tedious. In cost and accuracy comparisons, the total computing time to generate eigenvalues and eigenvectors, and then to perform the necessary exponential function evaluation, may be so large that various numerical differential equation solvers may have clear computational advantages. Some important considerations in the choice of a numerical integration method are stability and equation stiffness, CPU time, and the ease of varying the model parameters.

24B. A Three Compartment Thyroxine Model

Let us start with a fairly simple example [67], [65]. Thyroxine, when injected into the blood stream is then carried into the liver where it can be converted into iodine which is absorbed into the bile. Neither conversion nor absorption occurs instantaneously, so that some of the thyroxine that enters the liver is fed back to the blood before it can be converted. Hence the "multiple pass" compartmental model of Figure 24.1 is suggested by the physiology. If we let $x_i(t)$ be the concentration of thyroxine in compartment i at time $t \geq 0$, and assume that the rate of thyroxine transfer is proportional to its concentration in the donor compartment (an assumption which can be tested through the simulation process), then the mass balance equations flow model is $(t \geq 0)$

$$(24.1) \quad \begin{bmatrix} \dot{x}_1 \\ \dot{x}_2 \\ \dot{x}_3 \end{bmatrix} = \begin{bmatrix} -a_{21} & a_{12} & 0 \\ a_{21} & -a_{12}-a_{32} & 0 \\ 0 & a_{32} & 0 \end{bmatrix} \begin{bmatrix} x_1 \\ x_2 \\ x_3 \end{bmatrix} + b(t) .$$

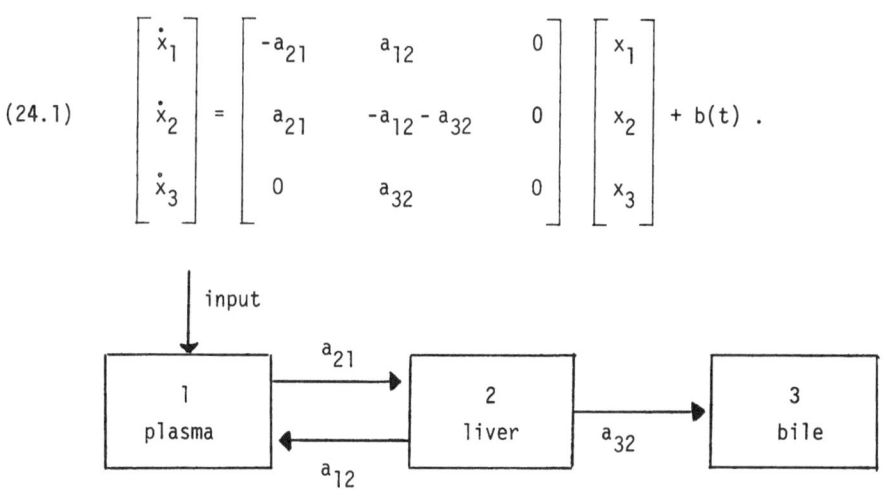

Figure 24.1.

The 3-compartment thyroxine model.

No absorption or transfer of thyroxine precisely at time zero supports the choice of initial conditions $x_2(0) = x_3(0) = 0$. Moreover we assume the "bolus input hypothesis" in which case $b(t) \equiv 0$ for all $t > 0$. We also will make the common assumption that the bolus input of thyroxine at $t = 0$ is normalized, i.e., $x_1(0) = 1$.

Correct simulation of this thyroxine model involves knowing or simply setting the values of the flow rates a_{ij}, initializing the levels of concentration of thyroxine in the different compartments (the selection of $x(0)$), and solving the differential equations in (24.1) numerically. Let us suppose that parameter estimation or discussions had led to the flux rates

$$(24.2) \qquad a_{21} = 0.05, \qquad a_{12} = 0.03, \qquad a_{32} = 0.01.$$

24C. Numerical Integration Methods and Some Inadequacies

In solving (24.1), probably the simplest numerical integration approach is to partition the time scale $[0, T]$ into equal sections of stepsize $h > 0$, compute $\dot{x}_i(0)$ from the model equations (24.1), and then utilize the elementary forward difference scheme (the Euler method)

$$(24.3) \qquad x^{(p+1)} = (I + hA)x^{(p)} \qquad\qquad p = 0, 1, 2, \ldots$$

with $x^{(0)} \equiv [1\ 0\ 0]^T$, $x^{(p)}$ corresponding to the actual solution value $x(t_p)$ where $t_p \equiv ph$, and where A is the compartmental matrix of (24.1).

Now suppose we were to use method (24.3) to solve the thyroxine model. The component-wise form of the difference equation is

$$
\begin{aligned}
x_1^{(p+1)} &= x_1^{(p)} + h(-\,a_{21}x_1^{(p)} + a_{12}x_2^{(p)}) \\
(24.4) \qquad x_2^{(p+1)} &= x_2^{(p)} + h(a_{21}x_1^{(p)} - a_{12}x_2^{(p)} - a_{32}x_2^{(p)}) \\
x_3^{(p+1)} &= x_3^{(p)} + h(a_{32}x_2^{(p)}) \ .
\end{aligned}
$$

Utilizing the a_{ij}-values of (24.2), and keeping the stepsize $h > 0$ quite small, Euler's method computes approximate values of the solution $x(t)$ at the discrete times $h, 2h, 3h, \ldots$.

Method (24.4) or (24.3) is a nice simple formula, but large truncation or formula error in Euler's method excludes it from being widely used in practice.

There are, however, available today many excellent methods for solving ordinary differential equation initial value problems which are much more accurate than the method of (24.3). Among the most popular and practical is the class of Runge-Kutta methods which are based on higher-order Taylor series expansions and assumptions on the number of function evaluations at each step of the process. The number q such that the accumulated truncation error $E_p \equiv x^{(p)} - x(t_p)$ is of order $O(h^q)$ is usually called the "order" of the method. For actual codes, see RKF45 in Forsythe et al. [98, p. 135] which is a fourth-fifth order Runge-Kutta-Fehlberg, or Shampine et al. [209], [208], or DVERK in the IMSL package [131] (examples of which were given in Section 23). The Runge-Kutta methods will almost always come up with an accurate solution and can be generally highly recommended in cases where the function evaluations are easy to perform and where the accuracy needed is small, about 10^{-4}. However, there are certain situations arising in biomathematical modeling and other areas where these methods are inadequate--circumstances either where the stepsize h should be chosen fairly large in value, or when the system of ordinary differential equations is "stiff". In these cases one must resort to special methods.

To demonstrate these two inadequacies, let us consider for the moment the scalar model

$$(24.5) \qquad \dot{x}(t) = f(x(t)), \qquad t \geq 0, \quad x(0) = 1.$$

A typical representative (sometimes called the "midpoint method") of the class of Runge-Kutta methods is the second order scheme [176, p. 82] which, for the model (24.5), has the formula

$$(24.6) \qquad x^{(p+1)} = x^{(p)} + hf[x^{(p)} + \frac{h}{2} f(x^{(p)})].$$

Thus with $f(x) \equiv ax$, $a < 0$, and $x^{(0)} \equiv 1$, say, we get

$$(24.7) \qquad x^{(p)} = (1 + ha + h^2 a^2/2)^p \qquad p = 0, 1, 2, \ldots .$$

The term in parentheses in (24.7) is greater than unity if the stepsize $h > -2/a > 0$. Hence if a large stepsize h is taken, then $x^{(p)}$ becomes indefinitely large. By contrast, the exact solution, $x(t) = \exp(at)$, decays since $a < 0$. Therefore the Runge-Kutta method (24.6) is unsuitable--that is, "numerically unstable"--for large h in the case of this model.

Let us also consider the concept of stiffness of the differential equation. The _time constant_ τ of the solution of the scalar equation (24.5) is the time it takes for the solution to decay by a factor of $1/e$. For instance, if $f(x) \equiv ax$, $a < 0$, in (24.5), then $x(t)$ decays by a factor of $1/e$ in time $\tau = - 1/a$. A stable ordinary differential equation is called _stiff_ when it has a decaying exponential solution with a time constant that is quite small compared to the time interval over which the equation is being solved. For the linear compartmental model $\dot{x}(t) = Ax(t)$, these decay rates are related to the relative magnitudes of the eigenvalues λ_i of A.

To illustrate this, we turn to the thyroxine model. The eigenvalues of the compartmental matrix A of (24.1) are

$$\lambda_1, \lambda_2 = \frac{1}{2} \left\{ - (a_{21} + a_{12} + a_{32}) \mp [(a_{21} + a_{12} + a_{32})^2 - 4a_{21}a_{32}]^{\frac{1}{2}} \right\}$$

$$\lambda_3 = 0 .$$

Thus if the values in (24.2) are used, and the model (24.1) is solved numerically at the times t = 1, 2, ..., 5 by a Runge-Kutta method (in this case using DVERK in the IMSL package), there are no difficulties because the eigenvalues of A are not spread too far apart: $\lambda_1 = - 8.40 \times 10^{-2}$, $\lambda_2 = - 5.95 \times 10^{-3}$. The program execution time on a CDC CYBER 72 is 0.017 CPU seconds. Now let us suppose that we simulate the movement of thyroxine by assuming that the flow rate of thyroxine from the blood stream to the liver is 100 times greater than the feedback rate of thyroxine from compartment two to compartment one (see Figure 24.1), and in turn that this backflow is also 100 times greater than the rate of absorption of thyroxine into the bile. We simulate this by assigning the values $a_{21} = 100$, $a_{12} = 1$, and $a_{32} = 0.01$. The nonzero eigenvalues of A are now $\lambda_1 \doteq - 101$ and $\lambda_2 \doteq - 9.90 \times 10^{-3}$, which is a sizeable spread in magnitudes. Here the execution time of DVERK is 0.168 CPU seconds, or about 10 times the previous run time. If we carry this process one step further and choose a_{21} = 1000, a_{12} = 1, and a_{32} = 0.001 in the model, then the nonzero eigenvalues of A are λ_1 = - 1001 and λ_2 = - 9.99 × 10^{-4}, and the DVERK program execution time jumps to 1.41 CPU seconds. It is at this point, after a relatively large amount of computing time has been spent on getting the solution to the desired accuracy, that we recognize that in the last two cases the differential equation system is stiff.

In doing problems of the type and complexity of (24.1), an execution time of between 0.01 and 0.02 CPU seconds is what one would expect from an efficient program. The problem with the Runge-Kutta scheme in the second and third cases above is that the step size h is automatically being made smaller and smaller because the transient part of the solution (the exponential associated with the largest magnitude eigenvalue λ_1), which has long since disappeared, does not allow an increase in h. Thus a great deal of time is spent computing the solution to the desired accuracy. This can become very expensive in time and money, especailly for more complicated models.

24D. Implicit Methods

We have just seen that there are ordinary differential equation numerical integration methods which possess a "conditional" stability in that the method will converge and work well only for relatively small stepsize h. Often, however, numerical simulation of compartmental models will be needed over long time periods (e.g., see [192]) and thus we wish to retain a reasonably large step h to keep computing costs down, yet maintaining good accuracy of solution. We have also seen that most standard methods are not well suited for solving stiff equations. Special methods which are effective for stiff equations are currently under intensive investigation. Probably the simplest such method which, as it turns out, treats both long time simulation and stiff compartmental models is the backward Euler method.

In this backward Euler scheme, the basic model $\dot{x}(t)$ = Ax(t) is replaced by the approximating equation

$$(x^{(p+1)} - x^{(p)})/h \sim \dot{x}^{(p+1)} = Ax^{(p+1)} .$$

Solving for $x^{(p+1)}$, we get

(24.8) $x^{(p+1)} = (I - hA)^{-1} x^{(p)}$

provided the matrix $I - hA$ is nonsingular for the stepsize h. This invertibility is treated in the next result.

Theorem 24.1.   Let A be a nonzero compartmental matrix. Then

(a) $I - hA$ is invertible for any $h > 0$;

(b) $(I - hA)^{-1} \geq 0$ for all $h > 0$.

Proof.   Define the matrix K by $K \equiv cI + A$, where $c \equiv \max_i |a_{ii}|$. This constant c must be positive since the matrix $A \equiv [a_{ij}]$ is nonzero and diagonally dominant. Now $K \geq 0$, meaning that each of its elements is nonnegative and at least one is strictly positive. The Perron root of K is $c + \lambda_n$, where λ_n is the (real) eigenvalue of A of smallest magnitude (see Section 12). For any scalar $s \neq c$, consider the matrix

(24.9) $sI - K = (s-c)I - A = (s-c)[I - \frac{1}{s-c} A]$.

According the Luenberger [153, p. 198], this matrix has an existing inverse which is a nonnegative matrix with at least one nonzero element if and only if the scalar s is greater than the Perron root $c + \lambda_n$. Suppose now that h is an arbitrary positive number. Then there is a scalar $s_0 > c$ such that $h = 1/(s_0 - c)$. Now $s_0 - c > 0 \geq \lambda_n$ and so $s_0 > c + \lambda_n$. Hence by (24.9)

$$\frac{1}{s_0 - c} [I - \frac{1}{s_0 - c} A]^{-1} = \frac{1}{h} [I - hA]^{-1}$$

exists and is a nontrivial nonnegative matrix. Thus $[I - hA]^{-1}$ exists and is non-negative for any $h > 0$.

                                                                              ///

Now that we know that $I - hA$ is always invertible if $h > 0$ and A is compartmental, let us investigate the stability of this backward difference scheme. Define

$F \equiv F(h) \equiv (I - hA)^{-1}$ for whatever stepsize $h > 0$ is chosen. Then the backward Euler algorithm can be written as

(24.10) $x^{(p+1)} = Fx^{(p)} + d$ $p = 0, 1, 2, \ldots$.

A difference equation of this form, where d is a constant vector, is said to be asymptotically stable provided the spectral radius $\rho(F) < 1$ for any stepsize $h > 0$ [176, p. 71]. This translates into the fact that if the algorithm is asymptotically stable, then for any $h > 0$, errors do not grow or become magnified as time increases. We can state the following result for compartmental matrices.

Theorem 24.2.   Suppose the compartmental system that is represented by the compartmental matrix A has no traps. Then the backward Euler method (24.8) is asymptotically stable.

Proof.   Let $h > 0$ be arbitrary, and let $\{\lambda_i\}$ denote the eigenvalues of A. If the system has no traps, then A is invertible (Theorem 13.2) and the smallest magnitude eigenvalue $\lambda_n < 0$. Then $Re(\lambda_i) < 0$ for all i (Theorem 12.2). The eigenvalues of I-hA are $1-h\lambda_i$ and hence $Re(1-h\lambda_i) > 1$ for all i. Thus

(24.11) $|1 - h\lambda_i| \geq |Re(1 - h\lambda_i)| > 1$.

The eigenvalues of $F \equiv (I - hA)^{-1}$ are $(1 - h\lambda_i)^{-1}$, and due to inequality (24.11), we have $\rho(F) < 1$, the desired result.
                                                                    ▨

The backward difference scheme (24.8) may have an advantage for certain problems if the inverse matrix $F = (I - hA)^{-1}$ can be explicitedly calculated. That is because the compartmental matrix A is often sparse, of a particular form (such as mammillary or catenary), or the dimension of A is not too large (e.g., see [192, p. 41]).

Exercise 24.1.   For the thyroxine model (24.1), show that the inverse of I - hA is (h > 0)

$$\frac{1}{\Delta} \begin{bmatrix} 1+h(a_{12} + a_{32}) & ha_{12} & 0 \\ ha_{21} & 1 + ha_{21} & 0 \\ h^2 a_{21} a_{32} & ha_{32}(1 + ha_{21}) & \Delta \end{bmatrix} \geq 0 \, ,$$

where $\Delta \equiv \Delta(h) = 1 + h(a_{12} + a_{21} + a_{32}) + h^2 a_{21} a_{32}$. Numerically solve model (24.1) using the forward scheme (24.3) and the backward scheme (24.8) with both small and large stepsizes $h > 0$, and different choices for a_{12}, a_{21}, and a_{32}. Compare results with those of a Runge-Kutta scheme such as DVERK.

There are more accurate and efficient methods than the backward Euler method (24.8) that have been proposed especially for integrating stiff systems. Apparently the most successful methods are implicit, like (24.8). This is still very much an area of active research. Stiff equations are relatively expensive to solve. In applied problems such as compartmental modeling, fundamental quantities are known inaccurately, and consequently high accuracy solutions are rarely meaningful. Accuracies of just a few digits are common requests [208].

One method which has been developed by Gear [105], [104], [208], is available in the IMSL package under the subroutine name DGEAR. As with DVERK, the Gear sub-routine has stepsize control and variable order, where the order is chosen so as to maximize the stepsize through a comparison of estimates of $x^{(p)}$, then selecting the step based on these estimates and the size of the associated local truncation error.

The thyroxine model serves as a good example of using subroutine DGEAR for stiff equations. A code follows for the case when $a_{21} = 100$, $a_{12} = 1$, and $a_{32} = 0.01$ (so that the system is slightly stiff).

```
PROGRAM GEAR
EXTERNAL FCN, FCNJ
INTEGER N, METH, MITER, INDEX, IWK(3), IER, K
REAL X(3), WK(42), H, T, TOL, TEND
N = 3
T = 0.
```

```
      X(1) = 1.
      X(2) = 0.
      X(3) = 0.
      TOL = .0001
      H = .0000001
      METH = 2
      MITER = 2
      INDEX = 1
      DO  10  I = 1, 5
      TEND = FLOAT(I)
      CALL DGEAR (N, FCN, FCNJ, T, H, X, TEND, TOL, METH, MITER, INDEX, IWK, WK, IER)
      PRINT 100, X(1), X(2), X(3)
100   FORMAT (1X, F10.5, 4X, F10.5, 4X, F10.5)
10    CONTINUE
      STOP
      END


      SUBROUTINE FCN(N, T, X, XDOT)
      INTEGER N
      REAL X(N), XDOT(N), T
      XDOT(1) = - 100.* X(1) + 1.*X(2)
      XDOT(2) = 100.*X(1) - 1.01*X(2)
      XDOT(3) = 0.01*X(2)
      RETURN
      END
      SUBROUTINE FCNJ (N, T, X, PD)
      INTEGER N
      REAL X(N), PD(N, N), T
      RETURN
      END
```

Computing approximate values of the solution $x(t)$ of (24.1) at times $t = 1, 2, \ldots,$ 5, we get the same answers as before with DVERK, but in 0.093 CPU seconds computing time instead of the previous 0.168 seconds. If we move on to the third case considered before, when $a_{21} = 1000$, $a_{12} = 1$, and $a_{32} = 0.001$, so that system (24.1) is stiff to a significant degree, the outputted numerical values for the solution are identical to those before, but computed with a much greater savings in time: 0.096 CPU seconds as compared to 1.41 seconds with the Runge-Kutta scheme DVERK.

24E. Determining Model Stiffness

For the system of _linear_ equations, $\dot{x} = Ax + b$, where A is a constant matrix, the stiffness of the problem is determined by the relative magnitudes of the eigenvalues of A. If the eigenvalues are negative (or have negative real parts) and widely separated, then the system is stiff and there should be expected some difficulty in solving it by ordinary methods. It would be good if the system could be recognized as being stiff in advance and _without_ having to compute the eigenvalues of A. For compartmental matrices, we have just such a method. As usual, let the eigenvalues λ_i of the $n \times n$ matrix A be ordered as $|\lambda_1| \geq |\lambda_2| \geq \ldots \geq |\lambda_n|$. We will define the _spread of the eigenvalues_ as $s(A) \equiv |\lambda_1/\lambda_n|$ if $\lambda_n \neq 0$, and use this as a measure of the degree of stiffness of $\dot{x} = Ax + b$. Now if L is a known lower bound for $|\lambda_1|$, and U is a known nonzero upper bound for $|\lambda_n|$, then

$$s(A) = \frac{|\lambda_1|}{|\lambda_n|} \geq \frac{L}{U}$$

and so L/U provides a computable lower bound for the spread $s(A)$. If A is compartmental, then we can select U from Theorem 12.8, for instance, as $U \equiv \min_i |a_{ii}|$. Recall that it is conjectured that the largest magnitude eigenvalue λ_1 of a compartmental matrix A satisfies $|\lambda_1| \geq \max_i |a_{ii}|$ (see (12.6)); we do know this inequality holds when A is symmetrizable (see (23.25)). Hence for a collection of compartmental matrices A, we have

$$(24.12) \qquad s(A) = \frac{|\lambda_1|}{|\lambda_n|} \geq \frac{\max_i |a_{ii}|}{\min_i |a_{ii}|} \, .$$

If there is any question about the previous lower bound for $|\lambda_1|$, we do always know that

(24.13) $|\lambda_1| \geq \sum_{i=1}^{n} |a_{ii}|/n$

(see (12.17)) for any compartmental matrix A, so this substitute L could be used instead of $\max_i |a_{ii}|$.

To illustrate these inequalities, suppose we were simulating the distribution of lipid soluble chemicals in mammilian tissues with a compartmental model $\dot{x} = Ax + b$ as in [192]. Using some of the parameter values that are assigned to fractional transfer coefficients in [192], we come up with the compartmental matrix

$$
A = \begin{bmatrix}
-10.59 & 0 & 0 & 0 & 0 & 0 & 0 & 36.0 \\
0.362 & -0.394 & 0 & 0 & 0 & 0 & 0 & 44.0 \\
0 & 0 & -55.8 & 0 & 0 & 0 & 0 & 31.0 \\
0 & 0 & 0 & -16.0 & 0 & 0 & 0 & 30.2 \\
0 & 0 & 0 & 0 & -0.082 & 0 & 0 & 31.6 \\
0 & 0 & 0 & 0 & 0 & -0.254 & 0 & 23.7 \\
0 & 0 & 0 & 0 & 0 & 0 & -0.08 & 5.89 \\
10.0 & 0.394 & 55.8 & 16.0 & 0.082 & 0.254 & 0.08 & -202.39
\end{bmatrix}
$$

For this matrix,

$$
s(A) \geq \frac{\max_i |a_{ii}|}{\min_i |a_{ii}|} = \frac{202.39}{0.08} \doteq 2530
$$

or

$$
s(A) \geq \frac{\sum_{i=1}^{8} |a_{ii}|/8}{\min |a_{ii}|} = \frac{35.70}{0.08} \doteq 446 ,
$$

indicating definite stiffness of $\dot{x} = Ax + b$ (the actual eigenvalue ratio is $|\lambda_1/\lambda_8| = |-217.36/-0.00114| \doteq 191,000$).

SECTION 25. IDENTIFICATION OF COMPARTMENT VOLUMES

25A. The Basic Single Exit Compartmental Model

Up to now, our discussion on the identification of the compartmental model $\dot{x} = Ax + b$ has centered on the determination of the fractional transfer coefficients a_{ij} of the compartmental matrix A. In this section we shall also be concerned with the determination of the <u>volume</u> or <u>size</u> $V_i > 0$ of the ith compartment, $i = 1, 2, \ldots, n$ (see Section 9). If known, these together with the a_{ij}'s constitute the complete identification of the steady-state parameters of an n-compartment time-invariant system of <u>physical</u> compartments in which the volumes of the compartments are of primary importance.

In what is probably the most common type of experiment, the <u>concentration</u> of tracer in one or more compartments is sampled over time. Because of this, we shall recall equations (9 .1),

$$(25.1) \qquad \dot{c} = (V^{-1}AV)c + V^{-1}b,$$

as the fundamental compartmental equations to be studied. Here V is the constant diagonal matrix $diag[V_1, V_2, \ldots, V_n]$ and the concentration $c_i(t)$ of tracer in the ith compartment at time $t \geq 0$ is related to V_i by

$$c_i(t) = x_i(t)/V_i \qquad\qquad i = 1, 2, \ldots, n$$

which gives the conversion transformation $x = Vc$ from equation $\dot{x} = Ax + b$ to (25.1).

We will restrict ourselves here to a problem frequently encountered in physiology--the intravenous injection of an isotope into one compartment at time zero and the sequential sampling of that same compartment to estimate the tracer concentration as a continuous function of time and then utilize this concentration function to provide information about the characteristics of the system through the determination of the parameters a_{ij} and V_k of the model. For definiteness, suppose compartment 1 receives a bolus injection initially with a known amount D of tracer material. Thus the tracer model becomes

$$\dot{x}(t) = Ax(t), \qquad t \geq 0$$

(25.2)

$$x(0) = [D, 0, \ldots, 0]^T$$

so that the tracer concentration follows the dynamics of

$$\dot{c}(t) = Mc(t), \qquad t \geq 0$$

(25.3)

$$c(0) = [D/V_1, 0, \ldots, 0]^T.$$

Here M is defined as $V^{-1}AV$ like in Section 9, which by components is

(25.4) $m_{ij} = a_{ij}V_j/V_i$ $i,j = 1, 2, \ldots, n.$

If the concentration of labeled material in compartment 1 is sampled at various times, we assume that the discrete data points have a continuous representation as a sum of n exponentials such as in (23.6),

(25.5) $c_1(t) = \sum_{i=1}^{n} A_i e^{\lambda_i t}, \qquad t \geq 0 ,$

with all the constants $A_1, \ldots, A_n$ and $\lambda_1, \ldots, \lambda_n$ real and known. We wish to deduce from this function $c_1(t)$ the parameters a_{ij} and V_k of the model if possible.

We restrict the model further by assuming that there are no traps in the system and that loss of material will occur only from the first compartment ($a_{01} > 0$, $a_{0j} = 0$ if $2 \leq j \leq n$). This gives an "almost closed" system [196] of general physiological interest where the first compartment is typically the blood plasma and the leakage to the outside is due primarily to the kidney. The system is now a <u>single exit compartmental (SEC) system</u> [4], [8], where A is a SEC matrix (see Sections 14, 15) in which the sum of the first column is - a_{01}, and all other column sums of A are zero.

25B. Readily Identifiable Parameters

Let us now consider what parameters can be immediately identified from the proposed experiment and resulting output $c_1(t)$. First observe that since $c_1(t)$ is known in (25.5) for all $t \geq 0$, then from (25.3)

$$(25.6) \qquad V_1 = D/c_1(0) = D \Big/ \sum_{i=1}^{n} A_i$$

is an observable parameter. Second, there are some special formulas that can be applied; since A is an SEC matrix, one such formula is (16.23) in the form

$$(25.7) \qquad a_{01} = D/\int_{0}^{\infty} x_1(t)dt = D/V_1 \int_{0}^{\infty} c_1(t)dt$$

from which a_{01} can now be estimated. Also a_{11} can be easily identified, for from (25.2),

$$V_1 \dot{c}_1(0) = \dot{x}_1(0) = \sum_{j=1}^{n} a_{1j} x_j(0) = a_{11} x_1(0) = a_{11} V_1 c_1(0),$$

which yields

$$(25.8) \qquad a_{11} = \dot{c}_1(0)/c_1(0).$$

25C. The Catenary Single Exit System

If in addition we assume certain configurations of compartments, then much more can be said about the identification of the a_{ij} and the sizes V_k. Chain-like structure of compartments--that is, an n-compartment catenary system--is common in physiological and pharmacological studies (see Figure 11.2). It has been shown that the identification of the fractional transfer coefficients a_{ij} has a unique solution in the case when the single exit is from the nth compartment [24]. Our present case, where the single exit of the system is from the first compartment, is mentioned by Rubinow [194, p. 137]. He concludes that the system is identifiable for all of the a_{ij} and V_k, but that due to the algebraic nature of the nonlinear equations to be solved, it may happen that there is more than one solution to the equations; thus local identifiability of all model parameters is established. The global identifiability of the a_{ij} has been determined [8], and this is to be discussed below. Also results on the compartment sizes will be presented. More recently, Delforge [79] demonstrates the unique identification of the coefficients a_{ij} regardless of the number of the compartment from which the single exit occurs, but does not discuss the volumes V_k.

Let us now introduce the $n \times n$ constant matrix $K \equiv AV$, or by components, $k_{ij} \equiv a_{ij}V_j$. Also let $k_{01} \equiv a_{01}V_1$.

Exercise 25.1.   Show that K is a compartmental matrix since A is compartmental.
From the definitions of K and M, we see that $M = V^{-1}K$ and that from (25.2),

$$\dot{x} = Ax = KV^{-1}x = Kc$$

which is often taken as the original mass balance equation commencing the study of
tracer kinetics.   In this equation, k_{ii} represents the rate at which unlabeled
material leaves compartment i in the steady-state, and $k_{ij}(i \neq j)$ represents the
rate at which unlabeled material is transferred from compartment j to compartment i.
Independent of any tracer experiment, the conservation of steady-state flux of
unlabeled material into and out of the ith compartment would be expressed as

$$\sum_{\substack{j=1 \\ j \neq i}}^{n} k_{ij} + J_i = -k_{ii}$$

where J_i is the material flux from the outside environment into the ith compartment.
If there is no net flux from the outside into the system, then all $J_i = 0$.

   For $i \neq j$, the fraction $r_{ij} \equiv k_{ij}/k_{ji}$ is called the partition coefficient
between compartments i and j, and sometimes carries the name Donnan ratio. Bright
[43] shows that for the n-compartment catenary system of this section, the volumes
V_k and the rate constants k_{ij} can be uniquely determined provided the partition
coefficients are known.

   We set up an approach to the identification problem a bit differently than
before. To determine the a_{ij} by this approach, the following technical result is
needed.

LEMMA 25.1.   If in (25.2), A is a catenary matrix, then we have

(25.9) $x_{i+1}^{(i)}(0) = a_{i+1,i} \, x_i^{(i-1)}(0)$ $i = 1, 2, \ldots, n-1$.

Proof.   The model $\dot{x} = Ax$ implies the time derivative form $x^{(i)} = Ax^{(i-1)}$.   Thus

$$(25.10) \quad x_{i+1}^{(i)}(0) = \sum_{j=1}^{n} a_{i+1,j} \, x_j^{(i-1)}(0) = \sum_{j=1}^{n} a_{i+1,j} \, V_j c_j^{(i-1)}(0).$$

Since the system is catenary (see Section 11), then the derivative $c_p^{(k)}(0) = 0$ if $k < p - 1$ [43, p. 78], [124, p. 127]. Thus the last sum in (25.10) terminates at $j = i$. Also the catenary matrix A is tridiagonal. Hence the last sum in (25.10) reduces to

$$a_{i+1,i} \, V_i c_i^{(i-1)}(0) = a_{i+1,i} \, x_i^{(i-1)}(0).$$

///

THEOREM 25.2.   If system (25.2) has catenary structure, then the matrix A is uniquely identifiable.

Proof.   We proceed by showing that the columns of A can be successively identified.

The system is catenary, and so the element a_{21} is found from the known combination $- a_{11} - a_{01}$. Thus the first column of A is uniquely identified.

We recall that since there are no traps in the system, then A is invertible (Theorem 13.2) and each element in the top row of A^{-1} is $-1/a_{01}$ (Exercise 14.3). Thus the first component equation in $A^{-1}\dot{x} = x$ implies, for all $t \geq 0$,

$$(25.11) \quad \sum_{j=1}^{n} V_j \dot{c}_j = \sum_{j=1}^{n} \dot{x}_j = - a_{01} x_1 .$$

Because $\dot{c}_j(0) = 0$ for $j \geq 3$, then equation (25.11) becomes

$$(25.12) \quad \dot{x}_1(0) + \dot{x}_2(0) = - a_{01} x_1(0).$$

But $x_1(0)$ and $\dot{x}_1(0)$ are known from V_1 and $c_1(t)$. Hence (25.12) determines $\dot{x}_2(0)$. From $\dot{x} = A\dot{x}$, we get

$$\ddot{x}_1(0) = a_{11}\dot{x}_1(0) + a_{12}\dot{x}_2(0)$$

and thus the parameter a_{12} is uniquely determined since we know $\ddot{x}_1(0) = V_1 \ddot{c}_1(0)$.

Now

$$\dot{x}_1(t) = a_{11}x_1(t) + a_{12}x_2(t)$$

yields the values of $x_2^{(i)}(0)$ for all $i \geq 2$. In the equation

(25.13) $\ddot{x}_2(0) = a_{21}\dot{x}_1(0) + a_{22}\dot{x}_2(0) + a_{23}\dot{x}_3(0)$,

the quantity $\dot{x}_3(0) = V_3\dot{c}_3(0) = 0$, and $a_{22} = -a_{12} - a_{32}$. Thus a_{32} can be uniquely computed from (25.13). Hence the second column of A is uniquely determined.

For the third column of A, we start by considering

(25.14) $\dddot{x}_2(0) = a_{21}\ddot{x}_1(0) + a_{22}\ddot{x}_2(0) + a_{23}\ddot{x}_3(0)$

in which we can replace $\ddot{x}_3(0)$ by $a_{32}\dot{x}_2(0)$ according to the above Lemma. Thus (25.14) determines a_{23} uniquely. From

$$\dot{x}_2 = a_{21}x_1 + a_{22}x_2 + a_{23}x_3$$

the derivative $x_3^{(i)}(0)$ is known for all $i \geq 2$. The last term of equation

(25.15) $\dddot{x}_3(0) = a_{32}\ddot{x}_2(0) + a_{33}\ddot{x}_3(0) + a_{34}\ddot{x}_4(0)$

contains the factor $\ddot{x}_4(0) = V_4\ddot{c}_4(0) = 0$. If $-a_{23} - a_{43}$ is substituted for a_{33}, then (25.15) determines a_{43}, and then a_{33} is given by $-a_{23} - a_{43}$. Hence the third column of A is uniquely determined.

The pattern of proof is now established and by mimicking the procedures of the last paragraph, all remaining columns of A can be successively and uniquely identified.

▨

25D. Estimation of Compartment Volumes

Let us now turn to the estimation of compartment volumes. Our model is just as before (except we do not necessarily assume catenary configuration of compartments). First, we will observe that it is possible to compute the total volume of distribution

$$V_T \equiv V_1 + V_2 + \ldots + V_n$$

of the n-compartment system under the additional assumption (25.17) below.

Define the positive vector $v \equiv [V_1, V_2, \ldots, V_n]^T$ and note that upon using the ℓ^1-vector norm $\| \cdot \|_1$, we can write

(25.16) $\qquad V_T = \sum_{i=1}^{n} |V_i| \equiv \| v \|_1 .$

Now let us <u>assume</u> that the influx of unlabeled material from the outside environment is received only by compartment 1. We then have the condition

(25.17) $\qquad 0 = \sum_{j=1}^{n} k_{ij} + J_i \qquad i = 1, 2, \ldots, n$

from before, but the $J_2 = J_3 = \ldots = J_n = 0$. Letting $k_{ij} = a_{ij}V_j$, equation (25.17) can be rewritten in the linear system form

(25.18) $\qquad \sum_{j=1}^{n} a_{ij}V_j = \begin{cases} - J_1 & i = 1 \\ \\ 0 & 2 \leq i \leq n. \end{cases}$

Let $e_1 \equiv [1, 0, \ldots, 0]^T$. Then (25.18) is the same as $Av = - J_1 e_1$. We recall that since the compartmental system has no traps, then $A^{-1} \equiv [a_{ij}^{(-1)}]$ exists and so then $v = - A^{-1}J_1 e_1$. Since J_1 is a positive scalar, and recalling (25.16), the calculation proceeds as

(25.19) $\qquad V_T = \| v \|_1 = \| A^{-1}J_1 e_1 \|_1 = J_1 \| A^{-1} e_1 \|_1 = J_1 \sum_{i=1}^{n} |a_{i1}^{(-1)}| .$

However, we know that $T \equiv - A^{-1} \geq 0$ (Theorem 13.1 or (13.10)), that the (i,j)-element τ_{ij} of T is mean residence time, and that the mean residence time of a tracer particle in the entire system, given that it is loaded in the first compartment, is

(25.20) $\qquad \mu_1 = \sum_{i=1}^{n} \tau_{i1} = - \sum_{i=1}^{n} a_{i1}^{(-1)}$

(see Section 14). Combining (25.19) and (25.20) produces $V_T = J_1 \mu_1$. This equation is a variation of the <u>Stewart-Hamilton Principle</u> that the volume of material in the system is the input rate multiplied by the mean transit time [43], [194]. Now if we add together all of the equations of condition (25.17) and use the fact that the first column sum of K is - k_{01} while all other column sums are zero, then we find that $J_1 = k_{01}$. We have verified the next statement.

THEOREM 25.3.   In our n-compartmental single exit model (25.2) and (25.3), suppose that the influx of unlabeled material from the outside environment is received only by compartment 1.   Then

(25.21) $V_T = k_{01}\mu_1$

is the total volume of distribution of the system.

In our model, the parameter k_{01} can be computed directly from knowledge of $c_1(t)$: utilizing (25.7),

$$k_{01} = a_{01}V_1 = \frac{D}{V_1 \int_0^\infty c_1(t)dt} \, V_1 \; .$$

If the integration of $c_1(t)$ is carried out using (25.5), then

(25.22) $k_{01} = \dfrac{D}{-\sum\limits_{i=1}^{n} A_i/\lambda_i} \; .$

Also μ_1 is computable from the output $c_1(t)$ of the experiment. For, from the moments of the distribution ϕ defined in Section 14, we will have

(25.23) $\mu_1 = \int_0^\infty tc_1(t)dt / \int_0^\infty c_1(t)dt = -\sum\limits_{i=1}^{n} \dfrac{A_i}{\lambda_i^2} \Big/ \sum\limits_{i=1}^{n} \dfrac{A_i}{\lambda_i} \; .$

Individual compartment volumes V_i can also be calculated for the model of this section. Again, we get results for certain configurations of compartments.

THEOREM 25.4.   Suppose the compartmental matrix A of (25.2) is catenary, and assume that the influx of unlabeled material from the outside environment is received only by the first compartment. Then

(25.24) $V_i = \dfrac{a_{i,i-1}}{a_{i-1,i}} V_{i-1}$ $i = 2, 3, \ldots, n.$

Proof. Due to the last theorem, the first equation of (25.17) is

(25.25) $0 = \sum_{j=1}^{n} a_{1j}V_j + J_1$.

Since A is catenary, then (25.25) reduces to

$$0 = a_{11}V_1 + a_{12}V_2 + J_1$$

with $J_1 = k_{01} = a_{01}V_1$. Solve this equation for the V_2-term to get expression
(25.24) when i = 2:

(25.26) $a_{12}V_2 = - a_{11}V_1 - a_{01}V_1 = a_{21}V_1$.

Ilow the second equation of (25.17) is

$$0 = \sum_{j=1}^{n} a_{2j}V_j$$

since by hypothesis, $J_2 = 0$. Because A is catenary, this equation only has the
terms

$$o = a_{21}V_1 + a_{22}V_2 + a_{23}V_3 .$$

We solve for $a_{23}V_3$ and simultaneously replace a_{22} by $- a_{12} - a_{32}$ to get

(25.27) $a_{23}V_3 = - a_{21}V_1 + a_{12}V_2 + a_{32}V_2$.

By (25.26), the first term of this right-hand side is $- a_{12}V_2$. Substitution of this
quantity into (25.27) yields the desired result (25.24) when i = 3. The proof is
now completed if one continues inductively with the above argument.

⧄

Exercise 25.2. Let the compartmental matrix A of (25.2) be mammillary, and assume
as before that the influx of unlabeled material from the outside environment is
received only by compartment 1. Show that

$$(25.28) \qquad V_i = \frac{a_{i1}}{a_{1i}} V_1 \qquad\qquad i = 2, 3, \ldots, n.$$

If the influx of unlabeled material from the outside only goes to the first compartment, then the total volume is $V_T = k_{01}\mu_1$ (Theorem 25.3). If A is mammillary, then we also know that the individual volumes can be computed as in (25.28). As will be seen here, we can derive $V_T = k_{01}\mu_1$ directly from (25.28) by a different method.

Theorem 25.5. Suppose our compartmental system (25.2) is mammillary, with the central compartment as number one (see Figure 11.3), and suppose conditions (25.28) hold for all compartment sizes. Then $V_T = k_{01}\mu_1$ is the total volume of distribution of the system.

Proof. For any $i \geq 2$, the (i, 1)-entry of $I = AA^{-1}$ is

$$0 = a_{i1}a_{11}^{(-1)} + a_{ii}a_{i1}^{(-1)} + \sum_{k \neq 1, i} a_{ik}a_{k1}^{(-1)}.$$

Here the summation term is zero since A is mammillary. Thus

$$(25.30) \qquad 0 = a_{i1}a_{11}^{(-1)} + a_{ii}a_{i1}^{(-1)} = (a_{i1}/-a_{01}) - a_{1i}a_{i1}^{(-1)}$$

due to Exercise 14.3 and because A is mammillary. Hence

$$V_T = V_1 + V_2 + \ldots + V_n$$

$$= V_1 + V_1 \sum_{i=2}^{n} a_{i1}/a_{1i}$$

$$= V_1[1 + a_{01} \sum_{i=2}^{n} - a_{i1}^{(-1)}]$$

$$= V_1[1 + a_{01}a_{11}^{(-1)} + a_{01}\mu_1]$$

$$= V_1[1 + (a_{01}/-a_{01}) + a_{01}\mu_1]$$

$$= a_{01}V_1\mu_1 = k_{01}\mu_1$$

where we have successively used (25.28),(25.30),(25.20), and Exercise 14.3. ▨

Exercise 25.3. Suppose a compartmental model is catenary and also is closed so that $k_{0i} = 0$ and $J_i = 0$ for all i. Investigate the relation between elements of K. What is the relation between elements of A and M?

25E. Creatinine Clearance Model

A good example demonstrating the above calculations is an early study conducted by Sapirstein et al. [205] (also discussed in [35], [194]) to determine the volume of distribution of creatinine in the tissues by analysis of the plasma disappearance curve of creatinine after a single intravenous injection of this substance. In these experiments ten anesthetized dogs were used. After a single rapid injection of $D = 2.0$ gm of creatinine at time zero into the blood plasma via a foreleg vein, blood samples were taken at five minute intervals for sixty minutes. The model used is our equations (25.2) and (25.3) with two compartments, in which the first compartment is plasma, and the second is extravascular and is simply referred to as "tissue" (see Figure 16.2). The plasma creatinine concentration $c_1(t)$ was resolved into the following sum of two exponentials by the curve-peeling technique of Section 23,

$$c_1(t) = A_1 e^{\lambda_1 t} + A_2 e^{\lambda_2 t}$$

(25.31) $A_1 = 0.188$ mg/ml $\lambda_1 = -0.0161$/min

$A_2 = 0.321$ mg/ml $\lambda_2 = -0.1105$/min

where the numerical values shown here are those for the third dog. Then, according to our derivations, $V_1 = D/(A_1 + A_2) = 3.929$ liters by (25.6). Formula (25.7) yields

$$a_{01} = \frac{-D}{V_1(A_1/\lambda_1 + A_2/\lambda_2)} = 0.03491/\text{min}.$$

From (25.8), we get

$$a_{11} = \frac{\lambda_1 A_1 + \lambda_2 A_2}{A_1 + A_2} = -0.07563$$

so that $a_{21} = - a_{11} - a_{01} = 0.0407/min$. Using $x_1(t) = V_1c_1(t)$, we now see that (25.12) yields $\dot{x}_2(0) = 0.08144$. Then the equation after (25.12) gives $a_{12} = 0.05098/min$. Since $a_{22} = - a_{12}$ in this case, the matrix A of (25.2) is now uniquely identified, as ensured by Theorem 25.2.

Besides finding the matrix A, we also wish to determine the volumes or compartment sizes. The parameter V_1 is already known. We can get V_2 either through the total volume estimate, or by $V_2 = a_{21}V_1/a_{12}$. In the first case, V_T is calculated by (25.24) for which we need k_{01} and μ_1. The excretion rate k_{01} (which is $G \times 10^{-3}$ where G is the glomerular filtration rate in the Sapirstein paper) is given by (25.22) as $a_{01}V_1 = 0.1372$. From (25.23), the mean residence time is $\mu_1 = 51.54$ minutes. Hence $V_T = k_{01}\mu_1 = 7.069$ liters, so that $V_2 = V_T - V_1 = 3.14$ liters.

25F. Shock Therapy

Another application of the general model of this section is concerned with the therapy of shock, in work that was done in the Department of Surgery at The University of Texas Health Science Center at Dallas [213]. In a person in hemorrhagic shock there is a low-flow state in vital organs and a loss in volumes of fluids. All responses to the reduction of fluid volumes eventually result in the initiation of compensatory mechanisms directed at correction of the low-flow state. One such compensation is the movement of extracellular fluid (ECF) into the circulation. In treating this problem, getting some measurement of the ECF volume is essential. Sulfur-35 labeled sodium sulfate turns out to be a satisfactory tracer for obtaining information such as the estimate of ECF volume. The human system has been modeled by equations (25.2) and (25.3) with a 3-compartment catenary structure in which the first compartment is blood plasma, the second is the nonplasma ECF, and the third is the intracellular fluid (ICF). With this model, a measurement of the ECF volume V_2 can be made via (25.27), and the total volume of distribution can be calculated as in (25.24).

The S^{35} curves tend to show a rapid exponential slope followed by a second slower exponential disappearance curve. Visually a change in the slope of the plotted curve revealing the presence of an interacting third compartment (ICF)

cannot be seen, if at all, until the elasped time is greater than 180 minutes. For
clinical reasons there is strong interest in reducing the time period over which
observations are taken. In particular, it was hoped that we could estimate the ECF
volume based on 10 minutes of data to within 10% of that estimated from 180 minutes
of data. It could not be done. A reduction to a period of 60 minutes was the best
that could be done. For these reasons, and especially because of the apparent lack
of cellular penetration during the specified time period, the system also has been
modeled as a two-compartment system in which the first compartment is the plasma
and the second is the nonplasma ECF.

25G. Bounds and Approximations on Compartment Volumes

It is possible in our model that A is uniquely identifiable, but the volumes
cannot be exactly determined. In such a case, we can at least derive computable
upper and lower bounds on the total volume V_T.

THEOREM 25.6. For the model of (25.2) and (25.3), we have bounds on the total
volume of distribution of

$$(25.32) \qquad k_{01} \, \| \, A \, \|_1^{-1} \le V_T \le k_{01} \, \| \, A^{-1} \, \|_1 \, .$$

Proof. Let $J \equiv [J_1, J_2, \ldots, J_n]^T$ for the elements J_i in equation (25.17). Then

$$
\begin{aligned}
\| \, J \, \|_1 &= \sum_{i=1}^{n} J_i = \sum_{i=1}^{n} \left(- \sum_{j=1}^{n} k_{ij} \right) \\
&= - \sum_{j=1}^{n} \sum_{i=1}^{n} k_{ij} = - \sum_{j=1}^{n} (- k_{0j}) = k_{01}
\end{aligned}
$$

(25.33)

where the last equality holds since the system is single exit from compartment one.
Changing k_{ij} to $a_{ij} V_j$, equations (25.17) can be written as the linear system
$Av = - J$ or $v = - A^{-1} J$. Then

$$V_T = \| \, v \, \|_1 = \| \, A^{-1} J \, \|_1 \le \| \, A^{-1} \, \|_1 \, \| \, J \, \|_1 = \| \, A^{-1} \, \|_1 \, k_{01}$$

where the final equality comes from (25.33). On the other hand, the same equation
Av = - J yields

$$\| A \|_1 V_T = \| A \|_1 \| v \|_1 \geq \| Av \|_1 = \| J \|_1 = k_{01} .$$

▨

As an example, in the Sapirstein model we have identified k_{01} = 0.1372 and

$$A = \begin{bmatrix} -0.07563 & 0.05098 \\ \\ 0.04070 & -0.05098 \end{bmatrix}$$

so that $\| A \|_1$ = 0.11633 and $\| A^{-1} \|_1$ = 71.100. Hence by (25.32), there would be
the bounds on the total volume of $1.18 \leq V_T \leq 9.75$.

In practice, there are certain values proposed by investigators to be used as
approximations for the total volume V_T of the type of compartmental system studied
in this section: nV_1, $k_{01}/|\lambda_n|$, and D/A_n. All can be estimated as soon as the
curve-fitting for $c_1(t)$ in (25.5) has been completed. The estimate nV_1 seems in
general to be an approximation arrived at primarily because V_1 can be readily found
in (25.6). One virtue of the second estimate, $k_{01}/|\lambda_n|$, is that it lies between the
bounds of (25.32), as does V_T. To see this, we recall that since λ_n is an eigenvalue of A, then its magnitude must be less than or equal to any matrix norm of A,
and so in particular, $|\lambda_n| \leq \| A \|_1$. Hence $k_{01}/|\lambda_n|$ is greater than or equal to the
lower bound of (25.32). By contrast, λ_n^{-1} is an eigenvalue of A^{-1}, and so a similar
argument yields $|\lambda_n^{-1}| \leq \| A^{-1} \|_1$ and then $k_{01}|\lambda_n^{-1}| \leq k_{01} \| A^{-1} \|_1$. The third
estimate, D/A_n, is close to $k_{01}/|\lambda_n|$ in the following sense. According to (25.22)

$$\frac{k_{01}}{-\lambda_n} = \frac{D/\lambda_n}{\sum\limits_{i=1}^{n} A_i/\lambda_i} .$$

If we multiply numerator and denominator of this expression by λ_n, and let λ_n
approach zero, there results $k_{01}/|\lambda_n| \approx D/A_n$ for small λ_n since we assume the
ordering $\lambda_1 \leq \lambda_2 \leq \ldots \leq \lambda_n < 0$ for our model. For the Sapirstein study, the three

approximations of V_T are

$$2V_1 = 7.858, \quad k_{01}/|\lambda_1| = 8.521, \quad D/A_1 = 10.63$$

(here the subscripts of A and λ are reversed to keep consistent with (25.31) and the Sapirstein paper). Note that the last value, $D/A_1 = 10.63$, falls outside the bounds established by (25.32). We recall that the calculation allowed by the analysis of the model gave $V_T = 7.069$.

Exercise 25.4.   Suppose the compartmental system of (25.2) and (25.3) is such that the influx of unlabeled material from the outside environment is received only by compartment one. Show that

$$\max_{1 \leq i \leq n} V_i \geq k_{01} \| A \|_\infty^{-1} .$$

THEOREM 25.7.   Let the conditions of Exercise 25.4 hold, and suppose the inverse of the system matrix satisfies

(25.34) $$\max_{1 \leq i \leq n} |a_{i1}^{(-1)}| = |a_{11}^{(-1)}| .$$

Then $\max\limits_{1 \leq i \leq n} V_i = V_1$.

Proof.   Under the conditions of the hypothesis, vector J is $[J_1, 0, \ldots, 0]^T = J_1 e_1$, and $J_1 = k_{01}$ as before. The linear algebraic system $Av = -J$ can be written in terms of $v = [V_1, V_2, \ldots, V_n]^T$ as

$$v = -A^{-1}J = -J_1 A^{-1} e_1 = -k_{01} A^{-1} e_1 .$$

Thus

$$\max_{1 \leq i \leq n} V_i = \| v \|_\infty = k_{01} \| A^{-1} e_1 \|_\infty$$

$$= k_{01} [\max_{1 \leq i \leq n} |a_{i1}^{(-1)}|] = k_{01} |a_{11}^{(-1)}| = V_1$$

where the last equality is valid since for our SEC system, any element of the first row of A^{-1} is $- 1/a_{01}$.

▨

Now we can see that the previously mentioned quantity nV_1 is an upper bound for V_T under the hypothesis of the above theorem. For,

$$(25.35) \qquad V_T = \sum_{i=1}^{n} V_i \le n(\max V_i) = nV_1 .$$

The Sapirstein model serves as an illustration. From the identification of A, we compute A^{-1} and find that its first column is $[-28.63, -22.86]^T$. Thus condition (25.34) is satisfied and therefore Theorem 25.7 leads us to the fact that $V_1 \ge V_2$ for this model. This is in agreement with our previous results. From (25.35), we have the bound $V_T \le 2V_1 = 7.858$, an improvement over that furnished by (25.32).

SECTION 26. A DISCRETE TIME STOCHASTIC MODEL OF A COMPARTMENTAL SYSTEM

26A. The Markov Chain Model

So far, our discussions have centered nearly entirely on the continuous time deterministic linear compartmental model $\dot{x} = Ax + b$. An alternative that will be presented in this section is a discrete time probabilistic model, illustrated in the context of a 4-compartment model of tracer transport through the hepatic system for the differentiation among the major categories of liver diseases.

We assume that our n-compartmental system is __closed__, so that $a_{0j} = 0$ for all j (open systems are discussed in [91]). Instead of the constant coefficient system $\dot{x}(t) = Ax(t)$, $x(0) = x_0$, considered before, we will develop a __finite Markov chain model__ as follows. The set of n __states__ will be the n compartments of the system. A tracer particle can be in only one state at an instant in time. Transfers of state by a tracer particle will be assumed to occur only at times 0, h, 2h, 3h, ... For ease, we will normalize the stepsize h to one. If at time k the process is in state i, then at time $k + 1$ it will be in state j with __transition probability__ p_{ij}. Note here that the order of subscripts has been reversed from that of a_{ji} so that we will keep within the same notation as that of the main references [9], [199], [153], [155] for this section.

The transition probabilities p_{ij} associated with the Markov chain are placed as elements in the $n \times n$ __stochastic matrix__ $P \equiv [p_{ij}]$, whose nonnegative entries have a sum along any row of one (since the system is closed). Let $y_i(k)$, $i = 1,2,...,n$, be the probability that the state at the kth step of the process will be i. The $n \times 1$ vector $y(k)$ with components $y_1(k)$, $y_2(k)$, ..., $y_n(k)$, gives the probabilities that the Markov process takes on specific values. With an initial state y_0 specified, the successive probability vectors are generated by the standard discrete time difference equation model

$$y(k+1)^T = y(k)^T P \qquad k = 0,1,2,...$$

(26.1)

$$y(0) = y_0.$$

The original compartmental model $\dot{x} = Ax$, $x(0) = x_0$, has the solution $x(t) = \exp(tA)x_0$. We would like to determine the time-invariant linear system of difference equations with the same solution $\exp(tA)x_0$. It is clear that the correct system of difference equations to choose is ($h \equiv 1$)

$$x(k+1)^T = x(k)^T[e^A]^T \qquad k = 0,1,2,\ldots$$

with solution $x(k) = (\exp A)^k x_0 = \exp(kA)x_0$. We already know that if A is a closed compartmental matrix, then $P = (\exp A)^T$ is a row stochastic matrix (see (14.4) and [91]). Thus we see that it is quite reasonable to envision a closed compartmental system modeled as a finite Markov chain. (For a general discussion of the equivalence of Markov models to certain system dynamics models, the reader should see [200].)

Exercise 26.1.   [9] Let $A \equiv [a_{ij}]$ be a closed compartmental matrix. The matrix exponential $\exp(hA)$ has the convergent representation

$$e^{hA} = I + hA + (hA)^2/2! + (hA)^3/3! + \ldots ,$$

and $[\exp(hA)]^T$ is a row stochastic matrix. Show that if the positive scalar h is chosen no larger than $1/|a_{ii}|$ for any i, then the "approximation" $P_1 \equiv (I + hA)^T$ of $[\exp(hA)]^T$ is also a row stochastic matrix.

26B. The Liver Disease Model

Let us now study the liver disease model. This model has been introduced previously in Section 16 (see Figure 16.4). In particular, it was demonstrated there that the continuous time model $\dot{x} = Ax$ has a uniquely identifiable system matrix A from the given experimental setup. The upcoming discussion is based on work done at The University of Texas Health Science Center at Dallas, and reported in [9], [199].

In this study we are concerned with a mathematical model and its analysis dealing with the clinical problem of differentiation between the major categories of obstructive jaundice in infancy. The model presented is of rose bengal transport through the biliary system, and is designed to provide access to an improved accuracy of medical diagnosis.

There are two general subdivisions of liver disease, hepatitis and biliary atresia. Hepatitis reflects inflammation of the liver cells, whereas biliary atresia represents a severe anatomic obstruction somewhere in the biliary duct system within the liver (intrahepatic) or within the larger ducts leading out of the liver (extrahepatic).

Infectious hepatitis is believed to be caused by a virus and is more likely to spread under unsanitary conditions. Serum hepatitus is caused by another virus usually transmitted by any insufficiently sterilized instruments that penetrate the skin.

Extrahepatic biliary atresia, if there is not complete anatomic maldevelopment, can sometimes be alleviated by surgery. On the other hand, an operation on a patient with hepatitus may prolong the illness and possibly cause permanent liver damage. Because symptoms of both conditions are not mutually exclusive, the problem of differential diagnosis is ever present.

There are many liver tests, none of which give a clear picture of liver pathophysiology. One method of testing is based on measuring the disappearance rate of an injected dye or a radioactive tracer which is removed from the blood by the liver. In 1923, fluorescent dye rose bengal was first reported as a useful agent in the determination of the functional state of the liver. Improved evaluation was made possible when rose bengal was combined with the radioactive isotope I^{131}. Radioactive rose bengal is taken up by the liver, excreted into the bile, and eventually appears in the urine and feces. It has proved to be a good indicator of biliary tract obstruction and is presently perhaps the most reliable test for differentiation between hepatitis and biliary atresia.

The physiological basis for the model centers on the transfer of bilirubin [112, p. 1026], which is formed upon the breakdown of red blood cells. The average life span of red blood cells is about 120 days, and when it dies its membrane ruptures releasing hemoglobin into the circulation. The released hemoglobin is engulfed by bodily defense mechanism cells. In this process of red blood cell

destruction, bilirubin is formed. Bilirubin attaches itself to albumin and is
transported throughout the circulatory system. This "unconjugated" bilirubin
eventually arrives in the liver where the largest percentage is used to form bilirubin glucuronide; a small spinoff forms bilirubin sulfate. These two are called
"conjugated" bilirubin, which is excreted into the bile although a small portion
returns to the plasma and is reabsorbed by the liver or excreted by the kidneys.
That excreted into the bile finally enters the intestine where it is converted into
urobilinogen and is mostly excreted from the body with the feces giving the feces
its normal dark color.

Bilirubin transport is especially important because it is the excess of
bilirubin in the blood and body tissues that causes jaundice, the yellow discoloration of the skin associated with hepatic disease. Also bilirubin transport forms
the basis for the rose bengal test. The bilirubin pathway is shown in Figure 26.1.
Assuming the backflow from the intestine to the blood circulation is reasonably
negligible, then the idealized pathway of the radioactive rose bengal tracer is
illustrated in Figure 16.4.

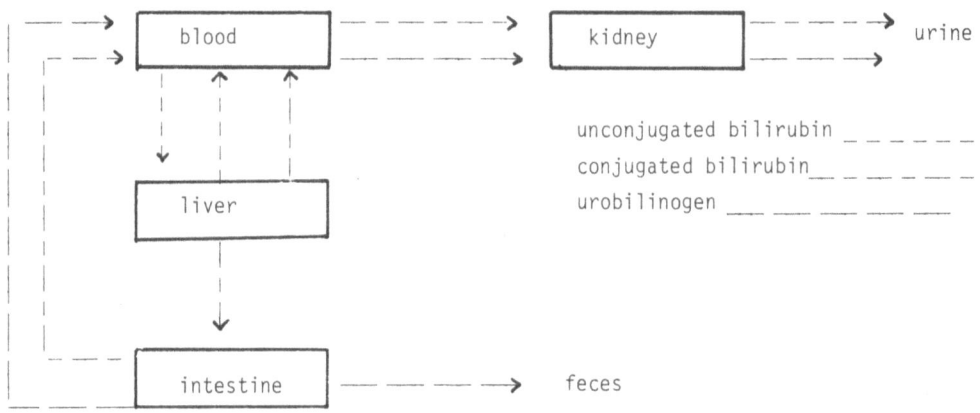

Figure 26.1.

Bilirubin Pathway

Previous attempts to mathematically model hepatic function have met with only
limited success. One reason is that only the rose bengal disappearance curve for

the blood compartment was used as system output. Up to now there has been no compartmental study which utilizes urine and fecal data. That may be due to the difficulty of obtaining samples. For example, in our work, over a 72-hour period, it
was only possible to take three or four urine and fecal specimens (see Figure 26.2).
Because of sums-of-exponentials sensitivity to statistical curve-fitting procedures,
these three or four data points are difficult to use. We cannot, with any degree of
confidence, determine the parameters of an exponential urine or fecal curve using
such a limited data base. One can attempt to bypass this problem by fitting the
model $\dot{x} = Ax$ directly to the data and thus avoid trying to fit three different
exponential curves (e.g., see program DIRFIT in Section 23), or, as we shall do in
this section, fit the model $y(k+1)^T = y(k)^T P$ directly to the data to determine the
transition probabilities p_{ij}.

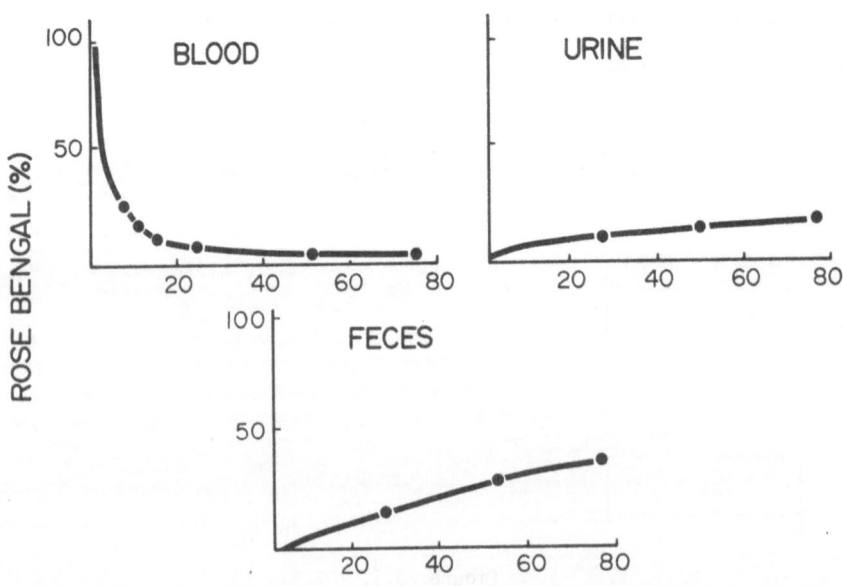

Figure 26.2

The basic goals of this study are to obtain information on the set of p_{ij}s for a patient with a particular liver disease, and to attain for certain liver conditions a characteristic set of p_{ij}s that can be used in the further diagnosis of abnormal liver function.

Following the pathway of the tracer as shown in Figure 16.4, the Markov chain model (26.1) has the four states

state 1 ≡ blood   state 3 ≡ liver

state 2 ≡ urine   state 4 ≡ feces.

In (26.1), the starting vector is

(26.2) $y_0 \equiv [1, 0, 0, 0]^T$

since a unit amount (100%) of rose bengal tracer is injected into the blood at time zero. Thus for any $k \geq 0$, we will think of $y_i(k)$, the ith component of the probability vector $y(k)$, as the amount (% remaining) of tracer in state i at time t = k. From Figure 16.4, we see that the form of the transition matrix $P \equiv [p_{ij}]$ in (26.1) is

(26.3) $P = \begin{array}{c c} & \begin{array}{cccc} 1 & 2 & 3 & 4 \end{array} \\ \begin{array}{c} 1 \\ 2 \\ 3 \\ 4 \end{array} & \left[\begin{array}{cccc} p_{11} & p_{12} & p_{13} & 0 \\ 0 & 1 & 0 & 0 \\ p_{31} & 0 & p_{33} & p_{34} \\ 0 & 0 & 0 & 1 \end{array} \right] \end{array}.$

Therefore the component equations of (26.1) are

$$y_1(k+1) = p_{11}y_1(k) + p_{31}y_3(k)$$

$$y_2(k+1) = p_{12}y_1(k) + y_2(k)$$

(26.4)   k = 0, 1, 2, ...

$$y_3(k+1) = p_{13}y_1(k) + p_{33}y_3(k)$$

$$y_4(k+1) = p_{34}y_3(k) + y_4(k) .$$

Note that since

(26.5) $p_{11} = 1 - p_{12} - p_{13}$ $p_{33} = 1 - p_{31} - p_{34}$

then there are only the four parameters, p_{12}, p_{13}, p_{31}, and p_{34}, to estimate from observations taken of states 1, 2, and 4.

26C. Simulation of the Hepatic System

Now that we have formulated our probabilistic model, which is (26.1) - (26.3), we can go through a step-by-step calculation, or simulation of the hepatic system. All that is needed are preselected parameter values p_{ij} conforming to

$$0 < p_{12}, p_{13}, p_{31}, p_{34} < 1,$$

$$0 < p_{12} + p_{13} < 1, \qquad 0 < p_{31} + p_{34} < 1.$$

For instance, if we assign $p_{12} = 0.03$, $p_{13} = 0.4$, $p_{31} = 0.1$, and $p_{34} = 0.1$, then (26.2) and (26.4) yield the graphical display in Figure 26.3 where time is measured in hours and the term "concentration" means the % of rose bengal remaining in the different states or compartments.

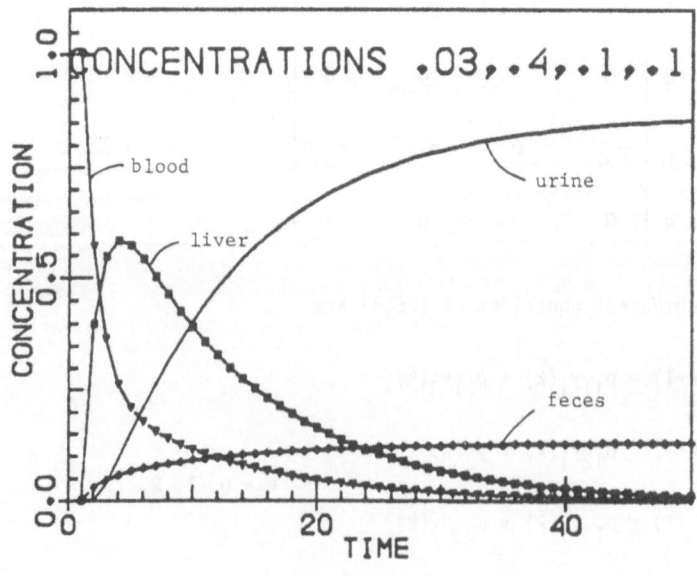

Figure 26.3.

Exercise 26.2. Carry out the simulation of the hepatic system with the p_{ij}-values assigned above. Does $y_1(k)$, $y_3(k) \to 0$ as $k \to \infty$? Does $y_2(k) + y_4(k) \to 1$ as $k \to \infty$? Why?

26D. Mathematical Analysis of the Model

Given a finite Markov chain, the ith state is said to be <u>absorbing</u> provided the transition probability $p_{ii} = 1$. Otherwise that state is nonabsorbing. The probabilities and states form an <u>absorbing Markov chain</u> if there is at least one absorbing state, and transition from each nonabsorbing state to some absorbing state is possible. Our model (26.1) - (26.3) is an absorbing Markov chain with absorbing states 2 and 4 (this is clear from Figure 16.4).

In the liver disease model, data are typically collected over a 72-hour period under the assumption that the basic time stepsize is $h = 1$ hour. Can we characterize what happens over the 72 steps to the probability vectors $y(0)$, $y(1)$, ..., $y(72)$? Our above simulation of the model suggests that the vector $y(k)$ approaches a limit as k gets large. There is a complete long-term behavior analysis that has been established for absorbing Markov chains, and we can use that to our advantage to answer questions of interest such as:

(a) What is the probability that a tracer particle is absorbed in state j given that it was in nonabsorbing state 1 at time zero?

(b) What is the expected number of times a tracer particle will be in nonabsorbing state j if it was in nonabsorbing state 1 at time zero and ran until it was absorbed?

(c) What is the expected number of transitions of state before a tracer particle is absorbed?

For the purpose of this long-term analysis, it is convenient to rearrange the order of the states so that the absorbing states are listed first. In our model, let us rename the states

$$1 \equiv \text{urine}, \quad 2 \equiv \text{feces}, \quad 3 \equiv \text{blood}, \quad 4 \equiv \text{liver},$$

so that the new transition matrix is the lower triangular block canonical form

$$(26.6) \quad C = \begin{array}{c} \\ u \\ f \\ b \\ 1 \end{array} \begin{bmatrix} \begin{array}{cc|cc} 1 & 0 & 0 & 0 \\ 0 & 1 & 0 & 0 \\ \hline P_{12} & 0 & P_{11} & P_{13} \\ 0 & P_{34} & P_{31} & P_{33} \end{array} \end{bmatrix} = \begin{bmatrix} \begin{array}{c|c} I & 0 \\ \hline R & Q \end{array} \end{bmatrix}$$

and the new form of the model (26.1) - (26.3) is

$$x(k+1)^T = x(k)^T C$$

$$(26.7) \qquad\qquad\qquad k = 0, 1, 2, \ldots .$$

$$x(0)^T = [0, 0, 1, 0]$$

Inductively we see that

$$(26.8) \qquad x(k)^T = x(0)^T C^k \qquad\qquad k = 1, 2, \ldots .$$

The reason for writing the absorbing Markov chain in this form is because the long-range behavior of the model is easy to visualize and analyze from the structure of C^k.

Exercise 26.3.   Show, for any positive integer k, that

$$(26.9) \qquad C^k = \begin{bmatrix} \begin{array}{c|c} I & 0 \\ \hline R_k & Q^k \end{array} \end{bmatrix}$$

where $R_k \equiv (I + Q + Q^2 + \ldots + Q^{k-1})R$.

We would expect in our model (26.7) that as time passes, the amount of rose bengal in the nonabsorbing states (3 $\equiv$ blood, and 4 $\equiv$ liver) would decrease since it is increasingly likely the rose bengal is absorbed into the urine or feces. Thus we would conjecture that $Q^k \to 0$ as $k \to \infty$, which is a standard result for general absorbing Markov chains.

Exercise 26.4. Prove that in (26.9) we have $Q^k \to 0$ as $k \to \infty$. Also show that $I - Q$ is invertible and, since $Q \geq 0$, then

$$(26.10) \qquad I + Q + Q^2 + Q^3 + \ldots = (I - Q)^{-1} \geq 0.$$

It is now clear that the limit matrix

$$(26.11) \qquad \lim_{k \to \infty} C^k = \begin{array}{c} \\ u \\ f \\ \\ b \\ 1 \end{array} \begin{array}{cc} u \quad f & b \quad 1 \\ \left[\begin{array}{c|c} I & 0 \\ \hline (I-Q)^{-1}R & 0 \end{array}\right] \end{array}$$

exists and that the long-range behavior of the model comes down to an analysis involving only $(I - Q)^{-1}$ and R. The matrix $M \equiv (I - Q)^{-1} \equiv [m_{ij}]$ in (26.10) is called the _fundamental matrix_ for the absolving Markov chain whose transition matrix is the canonical form C.

We are presently in a position to answer the questions (a) - (c) raised above, and thereby give a biological interpretation to some of these parameters. The answer to (a) comes from noting that since each C^k is a stochastic matrix, then the limit matrix of (26.11) is stochastic and so $(I - Q)^{-1}R$ is a matrix of probabilities. In particular, the probability that a tracer particle is absorbed in state j ($j=1,2$) given that it started in the blood state is

$$(26.12) \qquad [(I - Q)^{-1}R]_{1j}$$

(see [153, p. 241]).

For the answer to question (b), we again appeal to the standard theory of absorbing Markov chains [153, p. 240] to discover that if the process runs until absorption, then the quantity

$$(26.13) \qquad m_{1j} \equiv (1,j)\text{-element of } M$$

is the expected number of times (hours) a tracer particle will be in nonabsorbing state j ($j=1,2$) given that it began in the blood state.

By Theorem 2 of [153, p. 240], the answer to question (c) is

(26.14) $m_{11} + m_{12}$

that is, the sum (26.14) gives the expected number of times the process will stay in any of the nonabsorbing states before absorption given that it started in the blood.

Now let us define $F \equiv (I - Q)^{-1}R \equiv [f_{ij}]$ as the matrix in the lower left corner of (26.11). By (26.12), we have an interpretation for the first row entries f_{11} and f_{12}. Since F is a row stochastic matrix, then $f_{12} = 1 - f_{11}$. These interpretations can be expressed differently, as in the next exercise.

Exercise 26.5.   Given the model (26.7), show that

(a) the urine sequence $\{x_1(k)\}$ is strictly increasing, $\lim_{k \to \infty} x_1(k)$ exists, and the limit is f_{11};

(b) the fecal sequence $\{x_2(k)\}$ is strictly increasing, $\lim_{k \to \infty} x_2(k)$ exists, and the limit is $1 - f_{11}$;

(c) $\lim_{k \to \infty} x_3(k)$ and $\lim_{k \to \infty} x_4(k)$ both exist and are zero.

(It is also conjectured that the blood sequence $\{x_3(k)\}$ is strictly decreasing, but this is an open question that has not been answered at this time.)

Exercise 26.6.   Let $s \equiv p_{12}p_{31} + p_{12}p_{34}$ and $d \equiv s + p_{13}p_{34}$. Show that $f_{11} = s/d$ and $f_{21} = p_{12}p_{31}/d$. Clearly, $f_{11} > f_{21}$. Moreover, show that

$$
M = \begin{bmatrix} f_{11}/p_{12} & (1-f_{11})/p_{34} \\ f_{21}/p_{12} & (1-f_{21})/p_{34} \end{bmatrix}
$$

and verify that $m_{11} > m_{21}$, $m_{22} > m_{12}$. (These representations give all elements of F and M in terms of the basic four parameters p_{12}, p_{13}, p_{31}, and p_{34}.)

26E. Parameter Estimation

We must now discuss patient data analysis and the estimation of the basic four parameters p_{12}, p_{13}, p_{31}, and p_{34} for each of our patients. From among the group of patients, let us select just one, patient D, as an illustrative case. Consider the following data on the % of injected rose bengal remaining in the sampled body states for this patient (see Figure 26.2):

TABLE 26.1

Hour	Blood	Urine	Feces
4	0.3419		
12	0.2150		
24	0.1660	0.2063	0.0424
48	0.1121	0.3278	0.3308
72	0.0665	0.4346	0.4542

We wish to estimate the transition probabilities for patient D from this data using a least-squares method in the upcoming program LPEST. This program will require us to furnish initial estimates p_{12}^o, p_{13}^o, p_{31}^o, and p_{34}^o of these parameters. These starting guesses can be found as follows.

Step 1.   Between times 24 and 48 hours, the urine data appears to be approximately affine, with straight line curve $z(t) = 0.005063t + 0.08479$. Therefore if
we take $y_2^o(25) \equiv z(25) = 0.2114$, then from the second equation of (26.4), there
results $0.2114 = p_{12}(0.1660) + 0.2063$, from which we get $p_{12} = 0.03$. Carrying out
a similar process between times 48 and 72 hours yields $p_{12} = 0.04$. We select

(26.15) $p_{12}^o \equiv 0.035$,

the average of these two values.

Step 2.   From the first equation of (26.4), we have

$$y_1(1) = (1 - p_{12} - p_{13})y_1(0) + p_{31}y_3(0)$$

from which $p_{13} = 1 - p_{12} - y_1(1)$. The curvature of the blood data (draw a smooth
curve through the data points) suggests that $y_1^o(1) \equiv 0.7$ is a reasonable estimate of
$y_1(1)$. Then we assign

(26.16) $p_{13}^o \equiv 1 - p_{12}^o - y_1^o(1) = 0.26.$

Step 3. Equation four of (26.4) predicts that $y_4(1) = 0$. A linear inter-
polation between this value of zero and the 24th hour fecal data value 0.0424 yields
$y_4^o(2) = 0.002$. The third equation of (26.4) indicates that $y_3(1) = p_{13}$. Then
$y_4(2) = p_{34}y_3(1) = p_{34}p_{13}$. Thus

(26.17) $p_{34}^o \equiv y_4^o(2)/p_{13}^o \doteq 0.01$

is a reasonable initial choice for the parameter p_{34}.

Step 4. It remains to get an initial estimate of p_{31}. Since the compartmental system is closed, we know that the tracer model must satisfy

(26.18) $y_1(t) + y_2(t) + y_3(t) + y_4(t) = 1, \qquad t \geq 0 .$

Thus $y_3^o(24) = 0.5853$ and $y_3^o(48) = 0.2293$ from the data at 24 and 48 hours. The
first equation of (26.4) then gives us

(26.19) $y_1(25) = (1 - p_{12} - p_{13})y_1(24) + p_{31}y_3(24).$

Interpolating $y_1^o(25) = 0.161$ from the blood data curve, equation (26.19) yields

$$0.161 = (1 - p_{12}^o - p_{13}^o)(0.1660) + p_{31}(0.5853)$$

from which we obtain $p_{31} = 0.075$. Using (26.19) again, but with the 48 hour data,
yields $p_{31} = 0.135$. Hence we choose

(26.20) $p_{31}^o \equiv 0.10,$

the approximate average of these two, as the starting estimate for p_{31}.

We can now execute a least-squares estimation of the model parameters as
follows. At the beginning of the first hour we have, from (26.2),

(26.21) $y(1) \equiv [1, 0, 0, 0]^T .$

(Here, for the purpose of computing in FORTRAN, we increase each time by one and
therefore run the time index from 1 to 73 instead of from 0 to 72.) Starting with

(26.21), we successively compute all the difference equations of (26.4), then retain only those values corresponding to the hours at which we have observed patient data on the blood, urine, and feces. We then define the sum-of-squares function (here displayed for patient D)

$$F(p_{12}, p_{13}, p_{31}, p_{34}) = [0.3419 - y_1(5)]^2 + \ldots$$

(26.22)

$$+ [0.2063 - y_2(25)]^2 + \ldots + [0.0424 - y_4(25)]^2 + \ldots ,$$

and run the following code to compute the optimal p_{ij}-values. In this program, $X(1) \equiv p_{12}$, $X(2) \equiv p_{13}$, $X(3) \equiv p_{31}$, and $X(4) \equiv p_{34}$. The initial values calculated in the previous four steps are entered at the line DATA X, and the minimization sub-routine used is ZXMIN from the IMSL package [131] (subroutine ZXSSQ can also be used).

```
      PROGRAM LPEST
      EXTERNAL FUNCT
      DIMENSION X(4), H(10), G(4), W(12)
      N = 4
      NSIG = 4
      MAXFN = 1000
      IOPT = 0
      DATA X/.035, .26, .1, .01/
      CALL ZXMIN (FUNCT, N, NSIG, MAXFN, IOPT, X, H, G, F, W, IER)
      PRINT 9, X, F
9     FORMAT (F 8.4)
      END
      SUBROUTINE FUNCT (N, X, F)
      DIMENSION X(N), A(73), B(73), C(73), D(73)
      A(1) = 1.
      B(1) = 0.
      C(1) = 0.
      D(1) = 0.
```

```
        DO 5 I = 1, 72
        A(I+1) = A(I)* (1.-X(1) - X(2)) + X(3)*C(I)
        B(I+1) = B(I) + X(1)*A(I)
        C(I+1) = C(I)* (1. - X(3) - X(4)) + X(2)*A(I)
        D(I+1) = D(I) + X(4)*C(I)
5       CONTINUE
        F = (.3419 - A(5))**2 + (.2150 - A(13))**2
          + (.1660 - A(25))**2 + (.1121 - A(49))**2
          + (.0665 - A(73))**2 + (.2063 - B(25))**2
          + (.3278 - B(49))**2 + (.4346 - B(73))**2
          + (.0424 - D(25))**2 + (.3308 - D(49))**2
          + (.4542 - D(73))**2
        RETURN
        END
```

The output of program LPEST (patient D) is

$$(26.23) \quad \begin{array}{ll} \hat{p}_{12} = 0.0321 & \hat{p}_{31} = 0.1071 \\ \\ \hat{p}_{13} = 0.2619 & \hat{p}_{34} = 0.0154 \end{array}$$

with a sum of squares value at this point $\hat{p}$ of F = 0.0266. (It is always wise to check the final parameter values $\hat{p}_{ij}$ by running a simulation of the system using the model equations (26.4) with the estimates $\hat{p}_{ij}$ to see if the original experimental data is reconstructed.)

Note that we must also be ever watchful that the final computed p_{ij}-values fall within the feasible region

$$(26.24) \quad \begin{array}{l} 0 < p_{ij} < 1 \quad \text{for all } i, j \\ \\ 0 < p_{12} + p_{13} < 1, \quad 0 < p_{31} + p_{34} < 1. \end{array}$$

If there is any concern over producing parameter values not in the feasible region (26.24), then one can apply transformations to the constrained problem so that it becomes an unconstrained problem (see Chapter Z of [131], or Box [42]). The new version of IMSL carries subroutine ZXMWD which incorporates this transformation method to get a global minimum (under constraints) of a function of more than one variable.

According to our analysis of this system in Section 16 as a continuous time model of the form $\dot{x} = Ax$, we know that there is a unique determination of the compartmental matrix A. The same claim carries over to our model (26.1). To be confident that our method produces a unique set of optimal parameter values $\hat{p}_{ij}$ and that the $\hat{p}_{ij}$ do indeed furnish a global minimum of· F in the feasible region, random vectors of starting values can be used in LPEST. This has been done with the result that in each of the simulation trials, the final p_{ij}-values were identical to those of (26.23).

26F. Discussion

Parameter estimates $\hat{p}_{12}$, $\hat{p}_{13}$, $\hat{p}_{31}$, and $\hat{p}_{34}$ were computed using program LPEST for each of 14 patients in a control group given the 72-hour I^{131} rose bengal test and for whom there was already a **known diagnosis** of either atresia or hepatitis. Each infant patient was studied and treated at The Children's Medical Center in Dallas. Below is a sampling of the results for patients D. W. (neonatal hepatitis), M.R. (post-neonatal hepatitis), and L.M. (extrahepatic biliary atresia) [199] in Table 26.2. The post-hepatitis patient M.R. has a greater uptake of rose bengal by the liver, as shown by a larger p_{13}, as well as a smaller expected number of hours m_{11} that rose bengal remains in the blood, both indicating an improved liver function over, say, patient D.W. Although the mean times for hepatitis patient D.W. are more than for post-hepatitis patient M.R., these times are of the same order of magnitude when compared to the times of atresia patient L.M. Once the tracer particle is in the liver of patient L.M., it will stay for a considerable time as indicated by m_{12} = 949 hours. The results for L.M. indicate a possible outflow obstruction in the major biliary duct. This was indeed confirmed by an inspection of the patient. The lower p_{13}-value for patient L.M. indicates a slower uptake of the tracer by the liver. One might argue that a decreased uptake by the liver and not an outflow obstruction is the problem with this patient. Simulating an increased uptake by setting p_{13} = 0.55 instead of 0.25, holding all other p_{ij}-values the same as before, and recomputing M, it was found that the expected time spent by the tracer in the liver jumped to 1664 hours. Thus it would appear that the outflow blockage is in fact responsible for the long time that the rose bengal spends in the liver.

TABLE 26.2.

Patient	p_{12}	p_{13}	p_{31}	p_{34}	m_{11}	m_{12}	f_{11}	$1-f_{11}$
D.W.	0.0204	0.3423	0.0377	0.0101	10	77	0.22	0.78
M.R.	0.0523	0.5486	0.0415	0.0135	5	53	0.28	0.72
L.M.	0.0147	0.2583	0.0148	0.0002	55	949	0.81	0.19

There are also differences in the F matrices for these patients. For atresia patient L.M., 81% of the tracer is excreted in the urine as compared to 19% in the feces. This is the reverse of the two hepatitis patients, and again suggests a blockage in the major bile duct.

Other significant quantities to consider are ratios such as p_{31}/p_{34}, indicating the percentage of tracer reappearing in the blood from the liver as compared to the percentage going to the feces. Here there are large differences between atresia patient L.M. (p_{31}/p_{34} = 74) and hepatitis patients D.W. and M.R. (p_{31}/p_{34} = 3.7, 3.1, respectively). The large value for patient L.M. is a further indicator of a duct obstruction within that patient.

An investigation was made to uncover any possible patterns among the estimated parameter values that would lead to a method of obtaining a differentiation among the general physiological abnormalities of the hepatic system. It was observed that information of significance may possibly be obtained from the various ratios p_{ij}/p_{34}, as discussed above.

A standard statistical discriminant analysis program, BMD07M [88], was run on the data of 11 different parameters computed for each of the 14 patients. The purpose was to see if, in fact, the outputted parameter values could be used to properly classify each of the control group patients in their known diagnosis of atresia or hepatitis. The selection by the program of the first discriminator was the parameter ratio p_{31}/p_{34}. All patients but one were correctly classified; an atresia patient was incorrectly classified in the hepatitis group. On the next step, the initial variable, p_{31}/p_{34}, along with the second variable selected, p_{12}/p_{34}, succeeded in classifying all patients correctly in their known diagnosis subgroup.

In the future it is hoped that this model will provide a differential diagnosis prediction that can be sued, along with other information, to aid in clinical decisions that must be made when dealing with patients having abnormal liver function.

Second generation models, which attempt to improve on model (26.1), have been studied. One such investigation [198] suggests splitting the liver compartment into two interacting compartments: intrahepatic and extrahepatic. The goal here is to gain insight into differentiating between intrahepatic and extrahepatic atresia and intermediate cases. This model was not particularly successful, primarily due to doubt as to the uniqueness of the p_{ij}-values. Another variant of the 5-compartment model is a model which includes the intestine as a separate state or compartment (see Figure 26.1) [12]. Besides adding a compartment, a major deviation from the model techniques studied throughout this section was in the use of cubic splines to enrich the (necessarily) sparse data set obtained from the blood, urine, and feces states. Also a different approach for getting initial values for the minimization subroutine was employed. The results obtained through this model gave a performance index value that was of the same order of magnitude as the four-compartment model; at times the results were even somewhat better. However, unless there is some clear diagnostic advantage to the additional information given by this formulation, the fact that this method costs between ten and twenty times more than a run of the four-compartment model hinders its desirability. Nonetheless, the model did perform well, and therefore it does have potential as a diagnostic tool.

Other papers related to the model and mathematics of this section are Cobelli et al. [59], discussing human metabolism of bilirubin, and Bécus [21], who investigates the model $\dot{x} = A(z(t))x$, where $z(t)$ is a stochastic process. Discussion of Markov chains corresponding to compartmental models can be found in [238], [239]. Kinetics of radioactive rose bengal and bilirubin are investigated in [48], [107], [152], [230], [240].

SECTION 27. CLOSING REMARKS

The work cited in this text indicates the widespread use of compartmental systems in the biosciences. In many applications, first-order donor-controlled kinetics are assumed, giving rise to the standard linear compartmental equations (7.14) or (16.5). Probably the primary reason for the continuing interest in linear models is their mathematical simplicity and the fact that results can be determined analytically in closed form. Besides tracer kinetics, linear compartmental models are also usually adequate in the study of drug kinetics [47], [70]. In studies of metabolic systems [53], [54], [52], however, linear models sometimes unsatisfactorily describe these processes where certain nonlinear mechanisms need to be included. Chemical reaction kinetics, if the reactions are first-order, can be treated using linear compartmental models, but in general such kinetics give rise to nonlinear compartmental models. Linear compartmental analysis has had a large number of applications in the modeling of ecological systems [159], [113], [168]. The development of tracer experiments in medical research greatly stimulated the use of these models in ecology due to the analogy between the transport and measurement of tracer among components of an organism, and the transport and measurement of material in ecosystems. For an overview of linear compartmental modeling of ecological systems, see O'Neill [174].

This monograph has not attempted to treat linear time-dependent, nonlinear, stochastic, or time-delay compartmental models although some work has been done in these areas [14], [154], [201], [90], [85], [23], [51], [225], [224], [186], [221], [160], [202], [150], [142], [183].

Our efforts here have been directed towards reviewing some of the mathematical developments in compartmental analysis over the decade of the seventies and early eighties, and to present some new ideas and directions for further research. Special attention has been paid to particular compartmental models (Sections 2-5, 8, 26), structural properties of compartmental systems (Sections 11-15), and to the inverse problem, which is nicely outlined by Jacquez [136]. Of the three major phases of this problem (model specification, identifiability, and parameter estimation), we have concentrated on the last two. For the identifiability problem for linear time-invariant systems, three basic approaches have been proposed: (a) use of the transfer function for the experiment (Sections 16, 17, 18, 22); (b) similarity transformations and equivalent models (Sections 21, 26); (c) modal matrix/ component matrix approach (Section 20). In each of these cases the identifiability problem usually becomes a question of the existence of a solution to a system of nonlinear algebraic equations. (On this question of identifiability of biological systems, there has been a bit of confusion in the terminology which has lead to recent attempts to standardize definitions and concepts [58], [86], [146].) Each of the above mentioned areas merits additional research.

We have also observed that input and output reachability (connectability) are necessary conditions for the identifiability of our systems (Section 18, [75], [214]). It would be good if the additional conditions (18.19) and (18.20), which give the number of nonzero and nonredundant relations available from the transfer function, could be further refined.

Once a system has been found to be identifiable, we still must be concerned with the actual estimation of the model parameters (Sections 19, 23, 25). It has been seen that the equations discussed in Section 22 should be especially amenable to analysis due to their simple and appealing structure. Also the modal matrix or component matrix approach (Section 20) yields algebraic equations which, at worse, are quadratic and therefore should be more "easily" solvable by numerical means. In these cases further research needs to be done on both the theoretical nature of the problems, and on the development of convergent numerical algorithms to obtain the parameter estimates. Here sufficient conditions for identifiability could be simultaneously developed.

The component matrices serve double duty, being involved in the identification process as well as in measured parameter sensitivity ((20.25), [5]). A system may in theory be identifiable but have certain parameters which are so insensitive to variation for a given experiment that their values are poorly estimable. Additional work needs to be done on compartmental component matrices properties, other indicators of sensitivity such as the Hessian matrix in least-squares estimation, and on the general problem of design of experiments [136], [145], [84], [167], [83].

We have tried to intersperse examples and case studies throughout the text. However, there are many other illustrations of linear compartmental methods applied to both tracer and nontracer examples. A very small sampling is as follows: calcium metabolism in man [68]; residence time of thorium in man [34]; blood flow in cutaneous human tissue [207]; studies of polluted streams [115]; digital image processing [173]; first-order chemical reaction systems [241], [169, p. 251]; digestive processes of sheep [19], [39]. Other related books and articles which may be of general interest to the reader are [77], [130], [129], [217], [235], [216], [195], [106].

BIBLIOGRAPHY

1. R.C. Allen, Jr. and R.K. Wright, A spline method for recovering the matrix A in q´= Aq, given q in tabular form, available from The University of New Mexico, Albuquerque, N.M. 87106.

2. W.F. Ames, Canonical forms for nonlinear kinetic differential equations, Ind. Engr. Chem. Fundamentals 1 (1962) 214-218.

3. D.H. Anderson, Estimation and computation of the growth rate in Leslie's and Lotka's population models, Biometrics 31 (1975) 701-718.

4. D.H. Anderson, Iterative inversion of single exit compartmental matrices, Comput. Biol. Med.9 (1979) 317-330.

5. D.H. Anderson, Spectral sensitivity in linear biological models, J. Math. Biol., to appear.

6. D.H. Anderson, Structural properties of compartmental models, Math. Biosci. 58 (1982) 61-81.

7. D.H. Anderson, The inverse problem for differential equations involving unsampled components, to appear.

8. D.H. Anderson, The volume of distribution in single exit compartmental systems, Applied Nonlinear Analysis, V. Lakshmikantham (ed.), Academic Press, 1979, 425-437.

9. D.H. Anderson, J. Eisenfeld, S.I. Saffer, J.S. Reisch, and C.E. Mize, The mathematical analysis of a four-compartment stochastic model of rose bengal transport through the hepatic system, Nonlinear Systems and Applications, V. Lakshmikantham (ed.), Academic Press, New York, 1977, 353-371.

10. D.H. Anderson and E.C. Gartland, Jr., Identification of time series and compartmental models, Proceedings of the 10th IMACS World Congress, Vol. 2, 1982, 232-234.

11. D.H. Anderson and E.C. Gartland, Jr., The numerical solution of a nonlinear system arising in time series analysis, submitted to SIAM J. Sci. Stat. Comput., 1982.

12. D.H. Anderson and J.M. Mocenigo, A new five-compartment model of rose bengal transport through the hepatic system, with diagnostic applications, unpublished paper.

13. O.D.A. Anderson, An inequality with a time series application, J. Econometrics 2 (1974) 189-193.

14. G. Aronsson and R.B. Kellogg, On a differential equation arising from compartmental analysis, Math. Biosci. 38 (1978) 113-122.

15. K.J. Astrom and P. Eykhoff, System identification: a survey, Automatica 7 (1971) 123-162.

16. G.S. Atkins, Multicompartmental Models for Biological Systems, Methuen and Co., Ltd., London, 1969.

17. F. Ayres, Jr., Theory and Problems of Matrices, Schaum Publishing Co., New York, 1962.

18. H.T. Banks and G.M. Groome, Jr., Convergence theorems for parameter estimation by quasilinearization, J. Math. Analy. Appl. 42 (1973) 91-109.

19. E. Batschelet, Introduction to Mathematics for Life Scientists, Springer-Verlag, New York, 1976.

20. E. Batschelet, L. Brand and A. Steiner, On the kinetics of lead in the human body, J. Math. Biol. 8 (1979) 15-23.

21. G. A. Bécus, Random evolutions and stochastic compartments, Math. Biosci. 44 (1979) 241-254.

22. R. Bellman, Introduction to Matrix Analysis, McGraw-Hill, New York, 1960.

23. R. Bellman, Topics in pharmacokinetics - I: Concentration dependent rates, Math. Biosci. 6 (1970) 13-17.

24. R. Bellman and K.J. Astrom, On structural identifiability, Math. Biosci. 7 (1970) 329-339.

25. R. Bellman, J. Jacquez, R. Kalaba and S. Schwimmer, Quasilinearization and the estimation of chemical rate constants from raw kinetic data, Math. Biosci. 1 (1967) 71-76.

26. R. Bellman, H. Kagiwada and R. Kalaba, Inverse problems in ecology, J. Theoret. Biol. 11 (1966) 164-167.

27. R. Bellman and R. Kalaba, Quasilinearization and Nonlinear Boundary-Value Problems, American Elsevier, New York, 1965.

28. R. Bellman and R. Roth, A technique for the analysis of a broad class of biological systems, in Cybernetic Problems in Bionics, H.L. Oestreicher and D.R. Moore, eds., Gordon and Breach, New York, 1968, 725-737.

29. B.K. Berkstresser et al., Identification techniques for the renal function, Math. Biosci. 44 (1979) 157-165.

30. M. Berman, The formulation and testing of models, in Multicompartment analysis of tracer experiments, H.E. Hart, ed., Ann. N.Y. Acad. Sci. 108 (1963) 182-194.

31. M. Berman and R. Schoenfeld, Invariants in experimental data on linear kinetics and the formation of models, J. Appl. Phys. 27 (1956) 1361-1370.

32. M. Berman, E. Shahn and M.F. Weiss, The routine fitting of kinetic data to models, Biophys. J. 2 (1962) 275-287.

33. M. Berman and M.F. Weiss, SAAM Manual, U.S. Public Health Service Publication No. 1073, U.S. Govt. Printing Office, Washington, D.C., 1967.

34. S.R. Bernard, Estimates of microcurie-days residue, bone dose equivalent, and $(MPC)_w$ for thorium-232 in man using a mammillary model, Bull. Math. Biol. 35 (1973) 129-147.

35. S.R. Bernfeld, A model for detection of kidney diseases, unpublished paper.

36. H. Bernsten and J.G. Balchen, Identifiability of linear dynamical systems, Proc. IFAC Conf. Syst. Param. Estim., 3rd, Amsterdam, 1973, 871-874.

37. D.W. Bilheimer et al., Reduction in cholesterol and low density lipoprotein synthesis after portacaval shunt surgery in a patient with homozygous familial hypercholesterolemia, J. Clin. Invest. 56 (1975) 1420-1430.

38. G. Birkhoff and R.S. Varga, J. SIAM 6 (1958) 354-377.

39. K.L. Blaxter, N.M. Graham and F.W. Wainman, Some observations on the digestibility of food by sheep and on related problems, British J. Nutrition 10 (1956) 69-91.

40. E.K. Blum, Numerical Analysis and Computation, Theory and Practice, Addison-Wesley, Reading, Mass., 1972.

41. A. Bossi, C. Cobelli, L. Colussi and G. Romanin-Jacur, A method of writing symbolically the transfer matrix of a compartmental model, Math. Biosci. 43 (1979) 187-198.

42. M.J. Box, A comparison of several current optimization methods, and the use of transformations in constrained problems, Comput. J. 9 (1966) 67-77.

43. P.B. Bright, The volumes of some compartment systems with sampling and loss from one compartment, Bull. Math Biol. 35 (1973) 69-79.

44. R.W. Brockett, Finite Dimensional Linear Systems, Wiley, New York, N.Y., 1970.

45. R. Bronson, Matrix Methods, Academic Press, New York, 1969.

46. K.M. Brown, A quadratically convergent Newton-like method based upon Gaussian elimination, SIAM J. Numer. Analy. 6 (1969) 560-569.

47. R.F. Brown, Compartmental system analysis: state of the art, IEEE Trans. Biomed. Engr. BME-27 (1980) 1-11.

48. R.F. Brown and K.R. Godfrey, Problems of determinancy in compartmental modeling with application to bilirubin kinetics, Math Biosci. 40 (1978) 205-224.

49. J. Buell and R. Kalaba, Quasilinearization and the fitting of nonlinear models of drug metabolism to experimental kinetic data, Math. Biosci. 5 (1969) 121-132.

50. A.B. Callahan and S.M. Pizer, The applicability of Fourier transform analysis to biological compartmental models, in Natural Automata and Useful Simulations, H. Pattee, ed., Spartan, Washington, D.C., 1966.

51. M. Cardenas and J.H. Matis, On the time-dependent reversible stochastic compartmental model - II. A class of n-compartment systems, Bull. Math. Biol. 37 (1975) 555-564.

52. E.R. Carson, C. Cobelli and L. Finkelstein, The identification of metabolic systems - A review, presented at the Fifth IFAC Symp. Ident. Syst. Paramet. Estimat., Darmstadt, Germany, 1979.

53. E.R. Carson and E.A. Jones, Use of kinetic analysis and mathematical modeling in the study of metabolic pathways in vivo, Part I, N. Engl. J. Med. 300 (1979) 1016-1027.

54. E.R. Carson and E.A. Jones, Use of kinetic analysis and mathematical modeling in the study of metabolic pathways in vivo, Part II, N. Engl. J. Med. 300 (1979) 1078-1086.

55. H. Caswell, A general formula for the sensitivity of population growth rate to changes in life history parameters, Theoret. Pop. Biol. 14 (1978) 215-230.

56. J.P. Chander et al., A program for efficient integration of rate equations and least-square fitting of chemical reaction data, Comput. Biomed. Res. 5 (1972) 515-534.

57. C.T. Chen, Introduction to Linear System Theory, Holt, New York, 1970.

58. C. Cobelli and J.J. DiStefano, Parameter and structural identifiability concepts and ambiguities: a critical review and analysis, Amer. J. Physiol. 239 (1980) R7-R24.

59. C. Cobelli, M. Frezza and C. Tiribelli, Modeling, identification and parameter estimation of bilirubin kinetics in normal, hemolytic and Gilbert's states, Comput. Biomed. Res. 8 (1975) 522-537.

60. C. Cobelli, A. Lepschy and G. Romanin-Jacur, Comments on 'On the relationships between structural identifiability and the controllability, observability properties', IEEE Trans. Automat. Contr., AC-33 (1978) 965-967.

61. C. Cobelli, A. Lepschy and G. Romanin-Jacur, Identifiability of compartmental systems and related structural properties, Math. Biosci. 44 (1979) 1-18.

62. C. Cobelli, A. Lepschy and G. Romanin-Jacur, Identifiability results on some constrained compartmental systems, Math. Biosci. 47 (1979) 173-195.

63. C. Cobelli, A. Lepschy and G. Romanin-Jacur, Identification experiments and identifiability criteria for compartmental systems, in Compartmental Analysis of Ecosystem Models, J.H. Matis, B.C. Patten and G.C. White (eds.), International Co-operative Publishing House, Fairland, Md., 1979, 99-124.

64. C. Cobelli, A. Lepschy and G. Romanin-Jacur, Structural identifiability of linear compartmental models, Theoretical Systems Ecology, E. Halfon (ed.), Academic Press, New York, 1979, 237-258.

65. C. Cobelli and G. Romanin-Jacur, Controllability, observability and structural identifiability of multi-input and multi-output biological compartmental systems, IEEE Trans. Biomed. Engr. BME-23 (1976) 93-100.

66. C. Cobelli and G. Romanin-Jacur, On the structural identifiability of biological compartmental systems in a general input-output configuration, Math. Biosci. 30 (1976) 139-151.

67. C. Cobelli and G. Romanin-Jacur, Structural identifiability of strongly connected biological compartmental systems, Med. Biol. Engr. 13 (1975) 831-838.

68. S.H. Cohn, S.R. Bozzo, J.E. Jesseph, C. Constantinicles, D.R. Huene and E.A. Gusmano, Formulation and testing of a compartmental model for calcium metabolism in man, Radiat. Res. 26 (1965) 319-333.

69. D.A. Cook and G.S. Taylor, The use of the APL/360 system in pharmacology. A computer assisted analysis of efflux data, Comput. Biomedical Res. 4 (1971) 157-166.

70. D.J. Cutler, On the definition of the compartment concept in pharmacokinetics, J. Theoret. Biol. 73 (1978) 329-345.

71. G. Dahlquist, Problems related to the numerical treatment of stiff differential equations: a survey of recent work, ACM Proceedings of International Computing Symposium, 1973.

72. F. Daniels and E.H. Johnston, The termal decomposition of gaseous nitrogen pentoxide, J. Amer. Chem. Soc. 43 (1921) 53-71.

73. N. Davies, M.B. Pate and M.G. Frost, Maximum autocorrelations for moving average processes, Biometrika 61 (1974) 199-200.

74. R.H. Davis and J.H. Ottaway, The application of optimization procedures to tracer kinetic data, Math. Biosci. 13 (1972) 265.

75. E.J. Davison, Connectability and structural controllability of composite systems, Automatica 13 (1977) 109-123.

76. G. Debreu and I.N. Herstein, Nonnegative square matrices, Econometrica 21 (1953) 597-607.

77. J.G. Defares and I.N. Sneddon, An Introduction to the Mathematics of Medicine and Biology, North Holland Press, 1960, 533.

78. J. Delforge, A sufficient condition for identifiability of a linear system, Math. Biosci. 61 (1982) 17-28.

79. J. Delforge, An approach to the problem of linear system identifiability through the investigation of structural properties of the connection, injection and observation matrices, Proceedings of the Tenth IMACS World Congress, Vol. 2, 1982, 217-219.

80. J. Delforge, Necessary and sufficient structural condition for local identifiability of a system with linear compartments, Math. Biosci. 54 (1981) 159-180.

81. J. Delforge, New results on the problem of identifiability of a linear system, Math. Biosci. 52 (1980) 73-96.

82. J. Delforge, The problem of structural identifiability of a linear compartmental system: solved or not?, Math. Biosci. 36 (1977) 119-125.

83. J.J. DiStefano, Design and optimization of tracer experiments in physiology and medicine, Federation Proc. 39 (1980) 84-90.

84. J.J. DiStefano, Experiences with sequential optimal sampling schedule designs for pharmacokinetic and physiologic experiments, Proc. Tenth IMACS World Congress, Vol. 2, 1982, 203-205.

85. J.J. DiStefano, Tracer experiment design for unique identification of nonlinear physiological systems, Amer. J. Physiol. 230 (1976) 476-485.

86. J.J. DiStefano and C. Cobelli, On parameter and structural identifiability: nonunique observability/reconstructibility for identifiable systems, other ambiguities and new definitions, IEEE Trans. Auto. Contr. 25 (1980) 830-833.

87. J.J. DiStefano, A.R. Stubberud and I.J. Williams, Feedback and Control Systems, Schaum's Outline Series, McGraw-Hill Co., New York, 1967.

88. W.J. Dixon (editor), BMD Biomedical Computer Programs, second ed., University of California Press, 1968.

89. J. Eisenfeld, On identifiability of impulse-response in compartmental systems, Math. Biosci. 47 (1979) 15-33.

90. J. Eisenfeld, On washout in nonlinear compartmental systems, Math. Biosci. 58 (1982) 259-275.

91. J. Eisenfeld, Relationship between stochastic and differential models of compartmental systems, Math. Biosci. 43 (1979) 289-305.

92. J. Eisenfeld, Stochastic parameters in compartmental systems, Math. Biosci. 52 (1980) 261-275.

93. J. Eisenfeld, Survey of structural identification in compartmental systems, Proc. IEEE Internat. Sympos. Circuits Systems (Biomathematics-Biomedical Systems), 1980, 453-456.

94. J. Eisenfeld and B. Soni, Linear algebraic computational procedures for system identification problems, Proceedings of the First International Conference on Mathematical Modeling, ed. by X.J.R. Avula, Vol. 1, 1977, 551-561.

95. J.W. Evans, W.B. Gragg and R.J. LeVeque, On least squares exponential sum approximation with positive coefficients, Math. Comput. 34 (1980) 203-211.

96. D.K. Faddeev and V.N. Faddeeva, Computational Methods of Linear Algebra, Freeman and Co., San Francisco, 1963.

97. D. Fife, Which linear compartmental systems contain traps? Math. Biosci. 14 (1972) 311-315.

98. G.E. Forsythe, M.A. Malcolm and C.B. Moler, Computer Methods for Mathematical Computations, Prentice-Hall, Englewood Cliffs, N.J., 1977.

99. S.D. Foss, A method of exponential curve fitting by numerical integration, Biometrics 26 (1970) 815-821.

100. A.R. Frackelton, Jr. and J.K. Weltman, Diffusion control of CEA - binding to anti-CEA, to appear.

101. A. Froese and A.H. Sehon, Kinetics of antibody-hapten reactions, in Contemporary Topics in Molecular Immunology, Vol. 4, F.P. Inman and W.J. Mandy (eds.), Plenum Press, New York, 1975, 23.

102. D.G. Gardner, Resolution of multicomponent exponential decay curves using Fourier transform, Ann. N.Y. Acad. Sci. 108 (1963) 195.

103. D.G. Gardner, J.C. Gardner, G. Laush and W.W. Meinke, Method for the analysis of multicomponent exponential decay curves, J. Chem. Phys. 31 (1959) 978.

104. C.W. Gear, Numerical Initial Value Problems in Ordinary Differential Equations, Prentice-Hall, Englewood Cliffs, New Jersey, 1971.

105. C.W. Gear, The automatic integration of ordinary differential equations, Comm. ACM 14 (1971) 176-179.

106. General topics on "Multicompartment analysis of tracer experiments," Ann. N.Y. Acad. Sci. 108 (1963) 1-338.

107. H. Ghadim and A. Saas-Kortsak, Evaluation of the radioactive rose bengal test for the differential diagnosis of obstructive jaundice in infants, N. Engl. J. Med. 265 (1961) 351-358.

108. G.H. Golub and C. Reinsch, Singular value decomposition and linear systems solutions, Numer. Math. 14 (1970) 403-420.

109. M.S. Grewal and K. Glover, Identifiability of linear and nonlinear systems, IEEE Trans. Auto. Contr. AC 21 (1976) 833-837.

110. S.M. Grundy, H.Y. Mok, L.A. Zech and D. Steinberg, Transport of very low density lipoprotein triglycerides in varying degrees of obesity and hyper-triglyceridemia, J. Clin. Invest. 63 (1979) 1274-1283.

111. S.C. Gupta, Transform and State Variable Methods in Linear Systems, Krieger Publishing Co., Huntington, New York, 1971.

112. A.C. Guyton, Textbook of Medical Physiology, Saunders Co., Philadelphia, Pa., 1966.

113. E. Halfon (ed.), Theoretical Systems Ecology, Academic Press, New York, 1979.

114. F. Harary, Graph Theory, Addison-Wesley, Reading, Mass., 1969.

115. G.W. Harrison, A stream pollution model with intervals for the rate coefficients, unpublished paper.

116. C.M. Hart and R.J. Mulholland, Structural identifiability of compartmental systems based upon measurements of accumulated tracer in closed pools, Math. Biosci. 47 (1979) 239-253.

117. H.E. Hart, U. Malik and D. Sugarman, Determination of reaction rate constants in the mammillary system, Bull. Math. Biophys. 34 (1972) 87-92.

118. T.L. Hayden, D.W.A. Bourne and Y.T. Fu, Fourier transform analysis of radio-pharmaceutical data, to appear.

119. J.Z. Hearon, A monotonicity theorem for compartmental systems, Math. Biosci. 46 (1979) 293-300.

120. J.Z. Hearon, Residence times in compartmental systems and the moments of a certain distribution, Math. Biosci. 15 (1972) 69-77.

121. J.Z. Hearon, The kinetics of linear systems with special reference to periodic reactions, Bull. Math. Biophys. 15 (1953) 121-141.

122. J.Z. Hearon, The washout curve in tracer kinetics, Math. Biosci. 3 (1968) 31.

123. J.Z. Hearon, Theorems on linear systems, Ann. N.Y. Acad. Sci. 108 (1963) 36-68.

124. J.Z. Hearon and W. London, Path lengths and initial derivatives in arbitrary and Hessenberg compartmental systems, Math. Biosci. 14 (1972) 121-134.

125. I.N. Herstein, Topics in Algebra, Blaisdell, New York, 1964.

126. J.M. Hett and R.V. O'Neill, Systems analysis of the Aleut ecosystem, Artic Anthropol. 11 (1974) 31-40.

127. G. Hevesy, Radioactive Indicators. Their Application in Biochemistry, Animal Physiology and Pathology, Interscience Publishers, New York, 1948.

128. Y.C. Ho and B.H. Whalen, An approach to the identification and control of linear dynamic systems with unknown parameters, IEEE Trans. Auto. Contr. AC-8 (1963) 347.

129. D.J. Hoffman and S.K. Niyogi, Metal mutagens and carcinogens affect RNA synthesis rates in a distinct manner, Science 198 (1977) 513-514.

130. B. Horelick and S. Koont, Tracer methods in permeability, Unit 74, Undergraduate Mathematics and Its Applications (UMAP), 55 Chapel Street, Newton, Ma. 02160.

131. International Mathematical and Statistical Libraries, Inc., 7500 Bellaire Blvd., Houston, Texas 77036.

132. I. Isenberg, R.D. Dyson and R. Hansen, Studies on the Analysis of fluorescence decay data by the method of moments, Biophys. J. 13 (1973) 1090.

133. J.A. Jacquez, A First Course in Computing and Numerical Analysis, Addison-Wesley, Reading, Mass., 1974, 316.

134. J.A. Jacquez, Compartmental Analysis in Biology and Medicine, Elsevier, New York, 1972.

135. J.A. Jacquez, Compartmental models of biological systems: Linear and non-linear, Applied Nonlinear Analysis, V. Lakshmikantham (ed.), Academic Press, New York, 1979, 185-205.

136. J.A. Jacquez, The inverse problem for compartmental systems, Proc. Tenth IMACS World Congress, Vol. 2, 1982, 186-189.

137. R.E. Kalman, Mathematical description of linear dynamical systems, SIAM J. Control 1 (1963) 152-192.

138. R.E. Kalman, On the general theory of control systems, Proc. First IFAC Congress 1 (1960) 481-491.

139. D.W. Kammler, Least squares approximation of completely monotonic functions by sums of exponentials, SIAM J. Numer. Analy. 16 (1979) 801.

140. R.B. Kellogg and A.B. Stephens, Complex eigenvalues of a nonnegative matrix with a specified graph, Linear Algebra Appl. 20 (1978) 179-187.

141. R.L. Kodell and J.H. Matis, Estimating the rate constants in a two-compartment stochastic model, Biometrics 32 (1976) 377-400.

142. G.S. Ladde, Cellular systems - II. Stability of compartmental systems, Math. Biosci. 30 (1976) 1-21.

143. P. Lancaster, Theory of Matrices, Academic Press, New York, 1969.

144. C. Lanczos, Applied Analysis, Prentice-Hall, Englewood Cliffs, N.J., 1956.

145. E.M. Landaw, Optimal multicompartmental sampling designs for parameter estimation - practical aspects of the identification problem, Proc. Tenth IMACS World Congress, Vol. 2, 1982, 200-202.

146. Y. Lecourtier and E. Walter, Comments on "On parameter and structural identifiability: nonunique observability/reconstructibility for identifiable systems, other ambiguities and new definitions," IEEE Trans. Auto. Contr. 26 (1981) 800-801.

147. E.S. Lee, Quasilinearization and Invariant Imbedding, Academic Press, New York, 1968.

148. A. Lemaître and J.P. Malenge, An efficient method for multiexponential fitting with a computer, Comput. Biomedical Res. 4 (1971) 555-560.

149. R.J. Lermit, Numerical methods for identification of differential equations, SIAM J. Numer. Analy. 12 (1975) 488-500.

150. R.M. Lewis and B.D.O. Anderson, Insensitivity of a class of nonlinear compartmental systems to the introduction of arbitrary time delays, IEEE Trans. Circ. Syst., to appear.

151. C.C. Lin and L.A. Segel, Mathematics Applied to Deterministic Problems in the Natural Sciences, Macmillan Publishing Co., New York, 1974.

152. W.M. Lively, S.A. Szygenda and C.E. Mize, Modeling techniques for medical diagnosis I. Heuristics and learning programs in selected neonatal hepatic disease, Comput. Biomed. Res. 6 (1973) 393-410.

153. D.G. Luenberger, Introduction to Dynamic Systems, John Wiley and Sons, New York, 1979.

154. H. Maeda and S. Kodama, Qualitative analysis of a class of nonlinear compartmental systems: nonoscillation and asymptotic stability, Math. Biosci. 38 (1978) 35-44.

155. D.P. Maki and M. Thompson, Mathematical Models and Applications, Prentice-Hall, Englewood Cliffs, New Jersey, 1973.

156. P. Mancini and A. Pilo, A computer program for multiexponential fitting by the peeling method, Comput. Biomedical Res. 3 (1970) 1-14.

157. M. Marcus and H. Minc, A Survey of Matrix Theory and Matrix Inequalities, Prindle, Weber, and Schmidt, Boston, 1964.

158. J. Markham, D.L. Snyder and J.R. Cox, A numerical implementation of the maximum-likelihood method of parameter estimation for tracer-kinetic data, Math. Biosci. 28 (1976) 275-300.

159. J.H. Matis, B.C. Patten and G.C. White (eds.), Compartmental Analysis of Eco-system Models, International Co-operative Publishing House, Fairland, Md., 1979.

160. A. Mazanov, Stability of multi-pool models with lags, J. Theoret. Biol. 59 (1976) 429-442.

161. N.H. McClamroch, State Models of Dynamic Systems, Springer-Verlag, New York, 1980.

162. L.A. Metzler, Stability of multiple markets: The Hicks conditions, Econometrica 13 (1945) 277-292.

163. M. Milanese and G.P. Molino, Structural identifiability of compartmental models and pathophysiological information from the kinetics of drugs, Math. Biosci. 26 (1975) 175-190.

164. C.F. Minder and I. McMillan, Estimation of linear compartmental model parameters using marginal likelihood, Biometrics 33 (1977) 333-341.

165. C. Moler and C. Van Loan, Nineteen dubious ways to compute the exponential of a matrix, SIAM Rev. 20 (1978) 801-836.

166. C. Monot and J. Martin, Inverse problem for linear compartmental models, in Mathematical Models in Biology and Medicine, N.T.J. Bailey (ed.), North-Holland, Amsterdam, 1974, 49-70.

167. F. Mori and J.J. DiStefano, Optimal nonuniform sampling interval and test input design for identification of physiological systems from very limited data, IEEE Trans. Auto. Contr. 104 (1979) 893-899.

168. R.J. Mulholland, Analysis of linear compartment models for ecosystems, J. Theoret. Biol. 44 (1974) 105-116.

169. B. Noble, Applications of Undergraduate Mathematics in Engineering, Macmillan Co., New York, 1967.

170. B. Noble, Applied Linear Algebra, Prentice-Hall, Englewood Cliffs, New Jersey, 1969.

171. J.P. Norton, An investigation of the sources of nonuniqueness in deterministic identifiability, Math. Biosci. 60 (1982) 89-108.

172. J.P. Norton, Normal-mode identifiability analyses of linear compartmental systems in linear stages, Math. Biosci. 50 (1980) 95-115.

173. J. Nosil, P. McOrmond, Digital images, mathematical modeling and physiological parameters, SPIE Vol. 206 Recent and Future Developments in Medical Imaging II, 1979, 245-249.

174. R.V. O'Neill, A review of linear compartmental analysis in ecosystem science, in Compartmental Analysis of Ecosystem Models, J.H. Matis, B.C. Patten and G.C. White (eds.), International Co-operative Publishing House, Fairland, Md., 1979, 3-28.

175. O. Ore, Graphs and Their Uses, Random House, New York, 1963.

176. J.M. Ortega, Numerical Analysis: A Second Course, Academic Press, New York, 1972.

177. J.M. Ortega and W.C. Rheinboldt, Iterative Solution of Nonlinear Equations in Several Variables, Academic Press, New York, 1970.

178. L. Padulo and M.A. Arbib, System Theory, W.B. Saunders Co., Philadelphia, Pa., 1974.

179. D.H. Parsons, Biological problems involving sums of exponential functions of time: an improved method of calculation, Math. Biosci. 9 (1970) 37-47.

180. B.C. Patten and M. Witkamp, Systems analysis of ^{134}Cesium kinetics in terrestrial microcosms, Ecology 48 (1967) 813-824.

181. S.M. Pizer, A.B. Ashare, A.B. Callahan and G.L. Brownwell, Fourier transform analysis of tracer data, in Concepts and Models of Biomathematics, F. Heinmets (ed.), Marcel Dekker, New York, 1969, 105.

182. Y. Plusquellec, J.L. Steimer and J.F. Boisvieux, Can compartment models account for monotonicity in pharmacokinetics?, Math. Biosci. 58 (1982) 19-25.

183. H. Pohjanpalo, System identifiability based on the power series expansion of the solution, Math. Biosci. 41 (1978) 21-33.

184. S.W. Provencher, A Fourier method for the analysis of exponential decay curves, Biophys. J. 16 (1976) 27-41.

185. S.W. Provencher, An eigenfunction expansion method for the analysis of exponential decay curves, J. Chem. Phys. 64 (1976) 2772-2777.

186. P. Purdue, Stochastic theory of compartments, Bull. Math. Biol. 36 (1974) 305-309.

187. A. Rescigno, On some topological properties of the systems of compartments, Bull. Math. Biophys. 26 (1964) 31-38.

188. A. Rescigno and G. Segre, Drug and Tracer Kinetics, Blaisdell, Waltham, Ma., 1966.

189. A.W. Roberts and D.E. Varberg, Convex Functions, Academic Press, New York, 1973.

190. J.S. Robertson and S.H. Cohn, Use of an analog computer in studies of strontium and calcium metabolism in man, Ann. N.Y. Acad. Sci. 108 (1963) 122-127.

191. J.S. Robertson, D.C. Tosteson and J.L. Gamble, The determination of exchange rates in three-compartment steady-state closed systems through the use of tracers, J. Lab. Clin. Med. 49 (1957) 497-503.

192. S.E. Rodecap and F.T. Lindstrom, Numerical simulation of a compartmental system for the distribution of lipid soluble chemicals in mammilian tissues, Comput. Biol. Medicine 6 (1976) 33-51.

193. A.J. Roques, Determinants: a short program, the Two Year College Math. J. 10 (1979) 340-342.

194. S.I. Rubinow, Introduction to Mathematical Biology, John Wiley and Sons, New York, 1975.

195. S.I. Rubinow, Mathematical Problems in the Biological Sciences, CBMS Vol. 10, SIAM Publications, Philadelphia, Pa., 1973.

196. S.I. Rubinow, On closed or almost closed compartmental systems, Math. Biosci. 18 (1973) 245-253.

197. S.I. Rubinow and A. Winzer, Compartmental analysis: an inverse problem, Math. Biosci. 11 (1971) 203-247.

198. S.I. Saffer, P.L. Daniel and C.E. Mize, The comparison of a four-compartment and five-compartment model of rose bengal transport through the hepatic system, in Nonlinear Systems and Applications, V. Lakshmikantham (ed.), Academic Press, New York, 1977, 657-670.

199. S.I. Saffer, C.E. Mize, U.N. Bhat and S.A. Sygenda, Use of nonlinear programming and stochastic modeling in the medical evaluation of normal-abnormal liver function, IEEE Trans. Biomed. Engr., BME-23 (1976) 200-207.

200. K.E. Sahin, Equivalence of Markov models to a class of system dynamics models, IEEE Trans. Sys. Man Cyber. SMC-9 (1979) 398-402.

201. I.W. Sandberg, A note on the properties of compartmental systems, IEEE Trans. Circ. Syst. CAS-25 (1978) 379-380.

202. I.W. Sandberg, On the mathematical foundations of compartmental analysis in biology, medicine, and ecology, IEEE Trans. Circ. Syst. CAS-25 (1978) 273-279.

203. S. Sandberg, Structural identification in compartmental systems, unpublished Ph.D. dissertation, The University of Texas Health Science Center at Dallas, 1981.

204. S. Sandberg, D.H. Anderson and J. Eisenfeld, On identification of compartmental systems, Applied Nonlinear Analysis, V. Lakshmikantham (ed.), Academic Press, New York, 1979, 439-448.

205. L.A. Sapirstein et al., Amer. J. Physiol. 181 (1955) 330.

206. R. Schoenfeld, Linear network theory and tracer analysis, Ann. N.Y. Acad. Sci. 108 (1963) 69-91.

207. P. Serjsen, Blood flow in cutaneous tissue in man studied by washout of radioactive xenon, Circulation Res. 25 (1969) 215-229.

208. L.F. Shampine and C.W. Gear, A user's view of solving stiff ordinary differential equations, SIAM Rev. 21 (1979) 1-17.

209. L.F. Shampine, H.A. Watts and S.M. Davenport, Solving nonstiff ordinary differential equations - the state of the art, SIAM Rev. 18 (1976) 376-411.

210. C.W. Sheppard, Basic Principles of the Tracer Method, John Wiley and Sons, New York, 1962.

211. C.W. Sheppard and A.S. Householder, The mathematical basis of the interpretation of tracer experiments in a closed steady state system, J. Appl. Physics 22 (1951) 510-520.

212. C.W. Sheppard and W.R. Martin, Cation exchange between cells and plasma of mammalian blood, J. Gen. Physiol. 33 (1950) 703-722.

213. G.T. Shires and C.J. Carrico, Current status of the shock problem, in Current Problems in Surgery, M.M. Ravitch (ed.), Year Book Medical Publishers, Chicago, Ill., 1966.

214. D.D. Siljak, On reachability of dynamic systems, Int. J. Syst. Sci. 8 (1977) 321-338.

215. A. Silvers, R.S. Swenson, J.W. Farquhar and G.M. Reaven, Derivation of a three compartment model describing disappearance of plasma insulin - [131]I in man, J. Clin. Invest. 48 (1969) 1461.

216. W. Simon, Mathematical Techniques for Physiology and Medicine, Academic Press, New York, 1972, Chapter V.

217. B. Singer and S. Spilerman, The representation of social processes by Markov models, Amer. J. Sociol. 82 (1976) 1-54.

218. S.M. Skinner, R.E. Clark, N. Baker and R.A. Shipley, Complete solution of the 3-compartment model in steady-state after single injection of radioactive tracer, Amer. J. Physiol. 196 (1959) 238-244.

219. M.R. Smith, S. Cohn-Sfetcu and H.A. Buckmaster, Decomposition of multi-component exponential decays by spectral analytic techniques, Technometrics 18 (1976) 467.

220. A.K. Soloman, Equations for tracer experiments, J. Clin. Invest. 28 (1949) 1297-1307.

221. T.T. Soong and J.W. Dowdee, Pharmacokinetics with uncertainties in rate constants - III. The inverse problem, Math. Biosci. 19 (1974) 343-353.

222. G. Strang, Linear Algebra and Its Applications, Academic Press, New York 1969.

223. J. Swartz and H. Bremermann, Discussion of parameter estimation in biological modeling: algorithms for estimation and evaluation of the estimates, J. Math. Biol. 1 (1975) 241-257.

224. A.K. Thakur and A. Rescigno, On the stochastic theory of compartments: III. General time-dependent reversible systems, Bull. Math. Biol. 40 (1978) 237-246.

225. A.K. Thakur, A. Rescigno and D.E. Schafer, On the stochastic theory of compartments: II. Multi-compartment systems, Bull. Math. Biol. 35 (1973) 263-271.

226. C.D. Thron, Structure and kinetic behavior of linear multicompartment systems, Bull. Math. Biophys. 34 (1972) 277-291.

227. C.C. Travis, On structural identifiability, Proc. Internat. Conf. Nonlinear Phenomena in Math. Sci., V. Lakshmikantham (ed.), Academic Press, New York, N.Y., to appear.

228. C.C. Travis and G. Haddock, On structural identification, Math. Biosci. 56 (1981) 157-173.

229. C.E. Treanor, A method for numerical integration of coupled first order differential equations with greatly different time constants, Math. Comput. 20 (1966) 39.

230. G.L. Turco, F. Gehmi, G. Molino and G. Segre, The kinetics of I-131 rose bengal in normal and cirrhotic subjects studied by compartmental analysis and a digital computer, J. Lab. Cln. Med. 67 (1966) 983-993.

231. S. Vajda, Structural equivalence of linear systems and compartmental models, Math. Biosci. 55 (1981) 39-64.

232. H.D. Van Liew, Method of exponential peeling, J. Theor. Biol. 16 (1967) 43-53.

233. J.M. Varah, A spline least squares method for numerical parameter estimation in differential equations, SIAM J. Sci. Stat. Comput. 3 (1982) 28-46.

234. R.S. Varga, Matrix Iterative Analysis, Prentice-Hall, Englewood Cliffs, New Jersey, 1962.

235. V. Vemuri, Modeling of Complex Systems, Academic Press, New York, 1978, 225.

236. E. Walter, G. LeCardinal and P. Bertrand, On the identifiability of linear state systems, Math. Biosci. 31 (1976) 131-141.

237. E. Walter and Y. Lecourtier, Unidentifiable compartmental models: What to do? Math. Biosci. 56 (1981) 1-25.

238. G.G. Walter, A compartmental model of a marine ecosystem, in Compartmental Analysis of Ecosystem Models, J.H. Matis, B.C. Patten and G.C. White (eds.), International Co-operative Publishing House, Fairland, Md., 1979, 29-42.

239. G.G. Walter, Compartmental models, digraphs, and Markov chains, in Compartmental Analysis of Ecosystem Models, J.H. Matis, B.C. Patten and G.C. White (eds.), International Co-operative Publishing House, Fairland, Md., 1979, 295-310.

240. A.D. Waxman, P.A. Leins and J.K. Siemsen, In vivo dynamics studies of hepatocyte function: a computer method for the interpretation of rose bengal kinetics, Comput. Biomed. Res. 5 (1972) 1-13.

241. J. Wei and C.D. Prater, A new approach to first-order chemical reaction systems, Amer. Inst. Chem. Engr. J. 9 (1963) 77-81.

242. G. Wilson, Factorization of the covariance generating function of a pure moving average process; SIAM J. Numer. Analy. 6 (1969) 1-7.

243. B.H. Worsley and L.C. Lax, Selection of a numerical technique for analyzing experimental data of the decay type with special reference to the use of tracers in biological systems, Biochem. Biophys. Acta 59 (1962) 1-24.

244. R.M. Zazworsky and H.K. Knudsen, Controllability and observability of linear time-invariant compartmental models, IEEE Trans. Auto. Contr. AC-23 (1978) 872-877.

245. L.A. Zech, S.M. Grundy, D. Steinberg and M. Berman, Kinetic model for production and metabolism of very low density lipoprotein triglycerides, J. Clin. Invest. 63 (1979) 1262-1273.

Bio-mathematics

Managing Editor: S. A. Levin

Springer-Verlag
Berlin
Heidelberg
New York

Volume 8
A. T. Winfree

The Geometry of Biological Time

1979. 290 figures. XIV, 530 pages
ISBN 3-540-09373-7

The widespread appearance of periodic patterns in nature reveals that many living organisms are communities of biological clocks. This landmark text investigates, and explains in mathematical terms, periodic processes in living systems and in their non-living analogues. Its lively presentation (including many drawings), timely perspective and unique bibliography will make it rewarding reading for students and researchers in many disciplines.

Volume 9
W. J. Ewens

Mathematical Population Genetics

1979. 4 figures, 17 tables. XII, 325 pages
ISBN 3-540-09577-2

This graduate level monograph considers the mathematical theory of population genetics, emphasizing aspects relevant to evolutionary studies. It contains a definitive and comprehensive discussion of relevant areas with references to the essential literature. The sound presentation and excellent exposition make this book a standard for population geneticists interested in the mathematical foundations of their subject as well as for mathematicians involved with genetic evolutionary processes.

Volume 10
A. Okubo

Diffusion and Ecological Problems: Mathematical Models

1980. 114 figures, 6 tables. XIII, 254 pages
ISBN 3-540-09620-5

This is the first comprehensive book on mathematical models of diffusion in an ecological context. Directed towards applied mathematicians, physicists and biologists, it gives a sound, biologically oriented treatment of the mathematics and physics of diffusion.

Journal of

Mathematical Biology

ISSN 0303-6812

Title No. 285

Editorial Board:
H. T. Banks, Providence, RI; **H. J. Bremermann,** Berkeley,
CA; **J. D. Cowan,** Chicago, IL; **J. Gani,** Lexington, KY;
K. P. Hadeler (Managing Editor), Tübingen;
F. C. Hoppensteadt, Salt Lake City, UT; **S. A. Levin**
(Managing Editor), Ithaca, NY; **D. Ludwig,** Vancouver;
L. A. Segel, Rehovot; **D. Varjú,** Tübingen in cooperation
with a distinguished advisory board.

The **Journal of Mathematical Biology** publishes papers in
which mathematics leads to a better understanding of biological phenomena, mathematical papers inspired by biological research and papers which yield new experimental data
bearing on mathematical models. The scope is broad, both
mathematically and biologically and extends to relevant
interfaces with medicine, chemistry, physics, and sociology.
The editors aim to reach an audience of both mathematicians
and biologists.

Contents:

Springer-Verlag
Berlin
Heidelberg
New York

Subscription information and sample copy upon request